Caring About User Interface

S. Scott Strickland

Edited by:
Paige Adams Strickland

ISBN: 978-0-9899488-7-6

… and this is an industrial love song – Joe Walsh, lead in to "Life's Been Good"

This book would not be possible without support of and exposure to creative, talented and exceptionally caring educators, relatives, friends and colleagues. Also, many medical professional made it even possible to write this book and be on my third chance to live (or more), but that is a story for another book.

From my pre-career journey: Dennis Dupps, taught technical drawing, architectural drawing and introduced the engineering discipline. Joel Senter a professor of psychology at the University of Cincinnati sponsored my first musings on usability. Joel inspired others as evidenced here: https://www.beaconsociety.com/more-about-r-joel-senter.html.

From Milacron / Siemens: Perry Anderson, Larry Bellew, John Berenberg, Rick Busick, Ken Dejager, Dave Farmer, Bill Flere, Bill Fancher, Roger Hart, Brad Hochstetler, Jim Krietemeyer, Tom Kuhnell, Tom Lincoln, Carrie List, Bruce Minturn, John Terry, and many others. John Berenberg and John Terry saw enough potential in a new college grad to hire me. Jim Krietemeyer contributed encouragement, brainstorming collaboration, mentorship, and leadership through both my technical and managerial career phases.

From Intelligrated / Honeywell: (Ditto Rick and Ken from

above), Austin Honeycutt, Mike Bell, Will Hatfield, Brent Knab, Dan Kosht, Anthony Le, Jim McCarthy, Neelema Ramineni, Tim Rau, Rich Rennie, James Shrider, Dave Sprauer, Jason Tulino, Dan Wells, Tim Williams, Jesse Wilson and many others. Rich Rennie provided a great example of leadership and helped make continuing my career for several more eventful years possible.

From Home: wife Paige, daughters Jenny & Megan, granddaughter Finleigh, grandsons Reed & Noah and my feline office assistant Sapphire. Paige made this book possible by encouraging it, editing it and many other ways, too numerous to mention.

CONTENTS

Preface

Hello, I'm Macintosh. It sure is great to get out of that bag.
-- Apple Keynote (January 30, 1984)

People view their systems through the user interface lens we create. We succeed based on making people's lives better and that is why I am writing "Caring About User Interface". If you love building great user experiences, have a deep compassion for people you build them for, and seek out like-minded people, you will have an exceptional career. I hope this book helps and encourages you along your fulfilling path forward.

Coming up, stay tuned for a sneak peek into the book's contents, but first a journey in the "Wayback Machine" to the career that made this book possible.

Figure 1: Wayback Machine (Peabody's Improbable History) - Wikipedia.com

My career journey sprang from a favorable mix of the personal computer revolution, technology fascination and a lot of encouragement from family and friends. I enthusiastically kept up on Apollo era NASA technology (and still do now with Artemis). The space race inspired me and many colleagues to pursue careers in science, technology, engineering and math (STEM). Our generation enjoyed a "giant leap" of space related inventions most significantly including computers based on the integrated circuit (IC).

Apt with the times, I started college as an electrical engineering major with the end goal of designing personal electronics such as TV remote controls. However if we think of technology similarly to music, I discovered more passion for writing the music (software) than for designing the instruments (hardware). After making the switch to software, I graduated in 1984 from the University of Cincinnati with a degree in math and computer science.

My experiences as a fascinated corporate intern and computer consultant to my father's real estate appraisal business foretold my future career. I watched with compassion as people struggled with command-line-driven computers such as the Apple II and IBM PC.

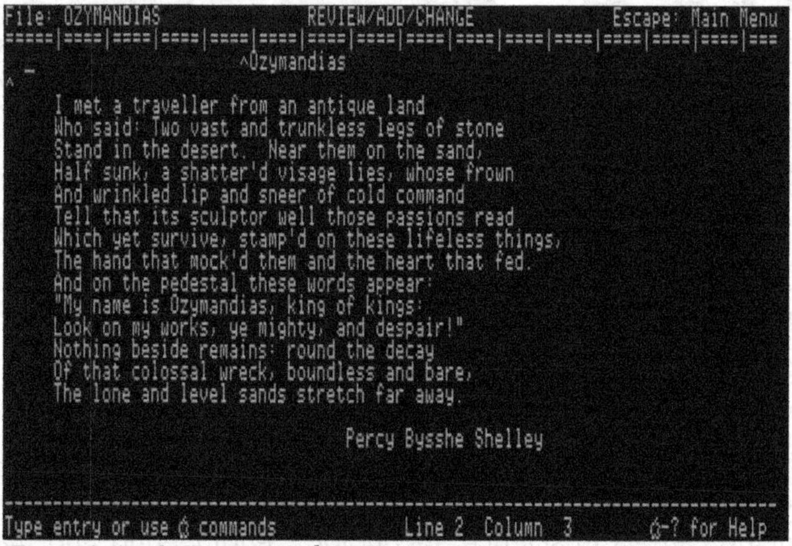

Figure 2: Apple II screen with rudimentary command driven interface - bytecellar.com/2016/06/05

Seeing the graphical user interface (GUI) of the Apple Lisa convinced me that personal computing was about to change radically and for the better. However, that promise took time. It would take both Apple's Macintosh and the advent of Microsoft Windows to adequately realize the GUI revolution.

Near the end of my college years, I approached Joel Senter, a professor of psychology at the University of Cincinnati and wrote a paper on the usability improvements offered by the

Macintosh's (now Mac's) GUI. Writing a paper about the Mac's breakthroughs as a first time Mac user was a meta experience that provided a decades-long career vision.

Figure 3: Macintosh: the computer that catalyzed my career - macrumors.com

In the nineties, I joined a team building one of the first machine tool GUIs and helped pioneer features for end user customization. That touch-screen-driven control system was my first realized GUI-driven product and certainly a career highlight.

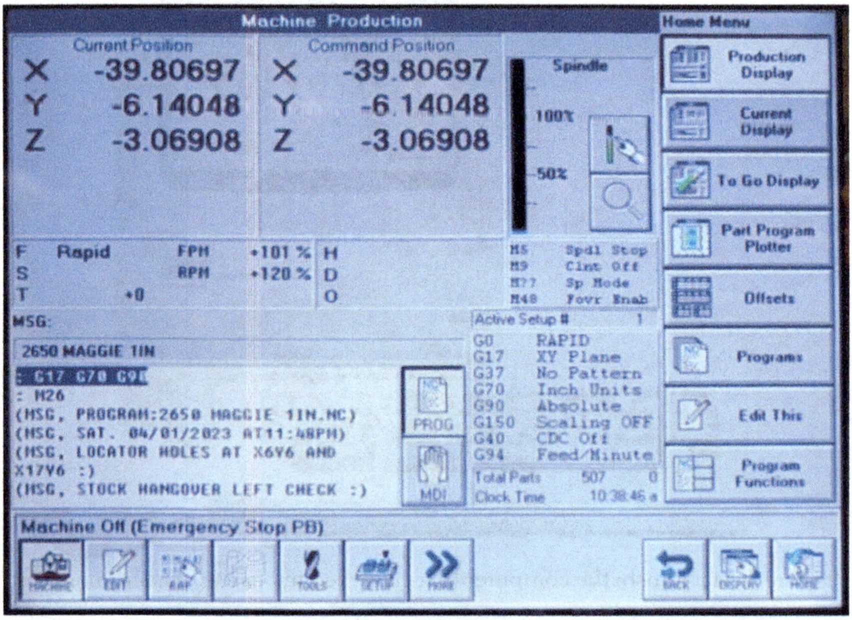

Figure 4: Our team's first touch screen Windows GUI. A dream realized - YouTube.com

The Internet supercharged the utility of GUI-based computing and set off unprecedented global team collaboration, radically changing how we live and work. "The Net" started as a modest communication medium, with email as its flagship. Combining Tim Berners-Lee's hypertext links with GUI browser apps transformed the Internet into "the Web" powerhouse we now frequent and largely take for granted.

For software developers, Web browsers like Internet Explorer and Google Chrome reduced distinctions between computers. Using the browser to hide differences between computer operating systems (an abstraction layer) freed developers from writing different software for different brands and devices. The cliche "Write once, test everywhere" meant browsers allowed writing code once, but still required testing on different browser software, devices and operating systems.

Happily, Mac/Windows/Linux platform independence had arrived. We eventually ran the same code on a workstation with mouse and keyboard that ran on a tablet (with some usability caveats). During this era, controlling a warehouse conveyor system with a multi-touch web application on an iPad (and Android tablet) became another career highlight.

Similarly, I witnessed the iPod, iPhone and Apple Watch bring about revolutionary personal consumer electronic devices with unique and groundbreaking specialized UIs at their core. Each UI was elegantly and intuitively matched to its device's size and environment.

For me, the breathtaking wonder of minuscule supercomputers and mind expanding variety of new applications never gets old. As I write this, we have virtual reality, augmented reality and spatial computing bringing new experiences and AI chatbots powering conversational UIs.

Those appreciating such breakthrough technologies imagine tremendous new potential. Increasingly, personal technology creates a future where the computer and device fade into the background. These increasingly seamless "user experiences" (UX) radically change our daily lives. With AI, speech, gestures, touch control, haptics (tactile feedback), projection-based augmented reality (AR), and many more, the choice of media and technologies promises unprecedented design opportunities.

In this book, I familiarize you with your audience, touch on design psychology, translate requirements into coherent products, point out pitfalls and steer towards an optimistic, forward-focused vision. I typically replace the term "user" with customer, people or audience to establish valued and caring partnerships. We should empower those we serve, not how "user" implies the other way around.

Shall we begin?

About this book

Welcome!

Some hints for approaching the contents of this book:

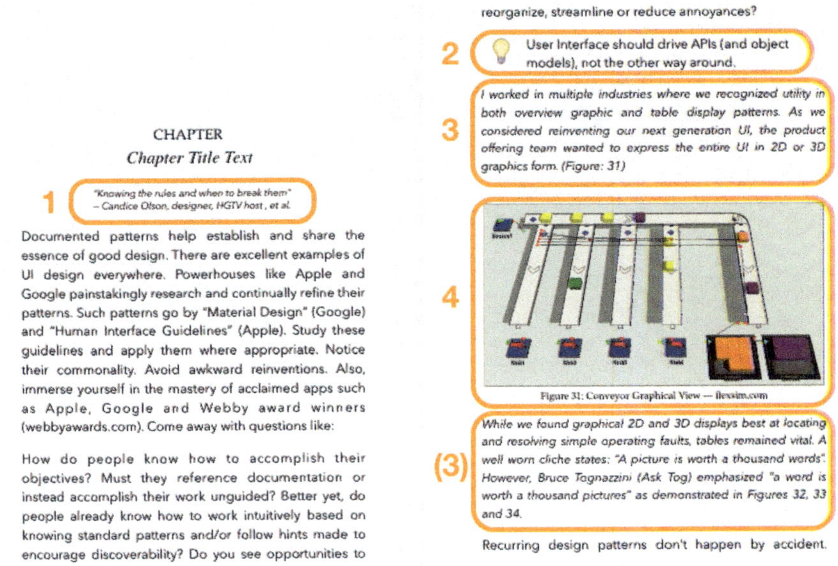

CHAPTER

Chapter Title Text

1 *"Knowing the rules and when to break them" – Candice Olson, designer, HGTV host , et al.*

Documented patterns help establish and share the essence of good design. There are excellent examples of UI design everywhere. Powerhouses like Apple and Google painstakingly research and continually refine their patterns. Such patterns go by "Material Design" (Google) and "Human Interface Guidelines" (Apple). Study these guidelines and apply them where appropriate. Notice their commonality. Avoid awkward reinventions. Also, immerse yourself in the mastery of acclaimed apps such as Apple, Google and Webby award winners (webbyawards.com). Come away with questions like:

How do people know how to accomplish their objectives? Must they reference documentation or instead accomplish their work unguided? Better yet, do people already know how to work intuitively based on knowing standard patterns and/or follow hints made to encourage discoverability? Do you see opportunities to reorganize, streamline or reduce annoyances?

2 User Interface should drive APIs (and object models), not the other way around.

3 *I worked in multiple industries where we recognized utility in both overview graphic and table display patterns. As we considered reinventing our next generation UI, the product offering team wanted to express the entire UI in 2D or 3D graphics form. (Figure: 31)*

4

Figure 31: Conveyor Graphical View — flexsim.com

(3) *While we found graphical 2D and 3D displays best at locating and resolving simple operating faults, tables remained vital. A well worn cliche states: "A picture is worth a thousand words". However, Bruce Tognazzini (Ask Tog) emphasized "a word is worth a thousand pictures" as demonstrated in Figures 32, 33 and 34.*

Recurring design patterns don't happen by accident.

1. Quote to set the tone.
2. Insights distill aha moments into takeaways for each chapter.
3. Personal experiences *presented in a smaller italic font.*
4. Illustrations of good and bad design to inspire you.

Part 1: Making a Great Product

CHAPTER ONE

Getting on Track

"People don't care how much you know until they know how much you care"
-- Teddy Roosevelt (maybe), 26th president, outdoorsman, veteran

The happy coincidence of the personal, Internet and mobile computing waves led to my career-long adventure designing and building user interface (UI) products. Perhaps as importantly, I've enjoyed and studied many of the inspiring products detailed herein.

Let's explore an experience where my team rescued an off-track project. We set the project back on track by delving deep into the customer's requirements and later ran into headwinds that we'll explore, too.

My memorable experience happened prior to writing any code. It was the Summer of 2022 during the pandemic. We collaborated with each other and our customers, almost all with online meetings.

In one such meeting, we gave a general-purpose Zoom demo to a prominent customer, and key stakeholders were openly skeptical about us delivering their vision. Their response to our demo was: "Where's our stuff?" Our customer wanted us to envision their UI rather than being treated to a phoned-in standard product demo. The specter of a lost golden opportunity loomed over our team.

Resolving to set things right, the sales team arranged a workshop to demonstrate our customer-specific concepts. Screenshots of their legacy UI guided us as to the customer's priorities. I asked many questions of both our project engineering team and of a customer representative. By stitching this information together, we understood our customer's UI requirements. Jesse, a member of our design team, mocked up graphical elements for crafting prototype high fidelity wireframes with Miro, an online collaboration tool.

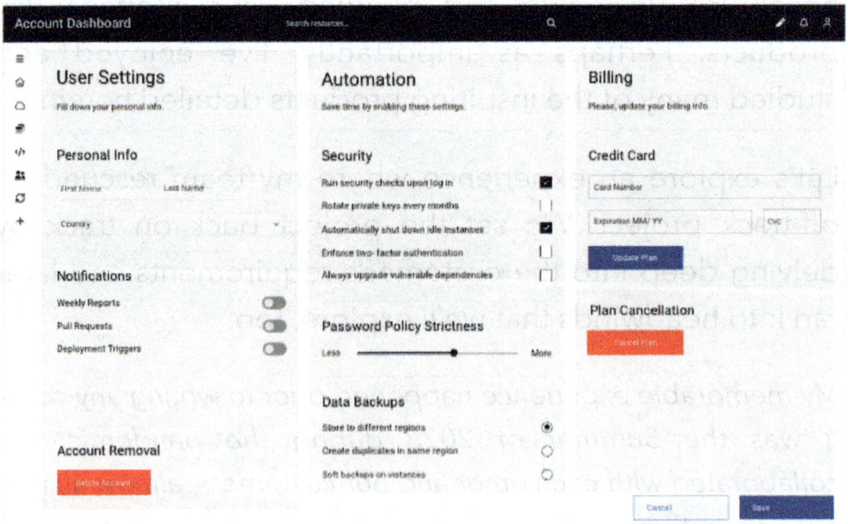

Figure 5: Example high fidelity wireframe — miro.com & author

Anyone in your team can build "wireframes" with Miro, PowerPoint, or Freeform that anchor productive design meetings.

With Miro, I spent several days practicing on-the-fly design changes and created wireframes based on our best understanding of how to improve on their familiar UI. Our team met internally to agree on the content. After our previous lackluster demo, the future of the project rode on this event.

The presentation meeting was a blur of customer team suggestions, juggling dashboard components, adding visual elements and taking notes at top speed. As facilitator, everything but the stream of ideas and the design vanished into the background much like the experience of a musician in tune with their audience. This experience "in the zone" stands as a career highlight for me.

After the meeting, customer response affirmed winning over hearts and minds. We cultivated the newly minted goodwill by further refining the design and updating the customer with the changes. Some notions of "we've always done it this way" needed dispelling. One takeaway for me: the UI would provide counts of top trending faults as a more useful alternative to industry standard routing lists. Summarized routing faults allowed our customer to quickly respond to high priority problems and maximize productivity of their site.

Enable people to make quick decisions with summarized data such as key performance indicators (KPIs) and real-time trends.

Figure 6: Examples of KPI Dashboards - dashboardbuilder.net

Our exacting customer received most of what they wanted, but more devils remained in the details. In hindsight, I regret not pushing harder to resolve remaining details with our team. One customer in a specific industry strongly influenced a more general purpose product design, possibly shortchanging others. We needed to make the right conclusions about custom versus standard. Annoyingly, the customer kept demanding changes even in the eleventh hour, though such ideas were not included in the original requirements. We were expected to include such new features for free. Regardless of some missteps, this peak experience illustrates close coordination and clear communication.

Agile processes typically dismiss design documentation. However, design validation must involve the customer, whether they validate requirements documentation, encourage continuous collaboration, or preferably, both. At the very least, a signed off requirements document avoids unpaid changes that consume profit. Change orders should handle missed details, not substitute for missing standard features.

> Detailed specifications and rigorous project managers control costs.

On the other hand, if you do not understand your customer's workflow and fail to validate key design choices, an otherwise capable product or feature may simply never get used (best case) or lead to even more catastrophic results. As an example of misunderstanding requirements, consider the following experience:

The Server Editor project failed to receive early design validation by project engineers (the primary customer). We either dismissed or misunderstood the target audience's preference to copy and tweak already working files. As designed, the tool required building its own inflexibly formatted files from scratch. This choice essentially barred project engineers from reading and editing existing files. The Server Editor worked as designed, but never connected with its target audience.

A younger colleague took on this challenge and invested many weeks on it. He thought the project was going well only to be disappointed by the result. Largely as a result of this experience, he moved on to an entirely different employer, (developing software for a lingerie retailer). I learned later that I inadvertently triggered his exit. I wanted to understand his interests and asked what he wanted to do next, but he interpreted my question as an invitation to move on to something more attractive.

All too often, we don't get quality in-depth access to customer expertise. Instead, we design in a vacuum knowing only hardware and software capabilities of the underlying system. For those of us with engineering backgrounds (like me), there is a temptation to design from the inside out using a personal knowledge of how the system works and a "mental model" of our own making. While preferable to ignorance, this practice is infamously known as "design by engineers for engineers".

Designing only from a mechanism-centric point of view (bottom up) yields bad results. The perspective of people using your software is far more important. To determine what that is, you must ask questions such as:

- Who uses this feature? Is it associated with a specific role such as maintenance?
- Who should have access to this feature and who shouldn't?
- How often is this feature used? Is it used in a certain part of the day or during setup?
- How will the software improve or replace existing processes?

- Are there any common/repetitive patterns to using this feature?
- What default values can I provide?

Regardless of whether or not you collaborate with a customer, team collaboration early in a project should involve people talented in imagining and verbalizing their ideas. Brainstorming in small teams energizes product creation, especially when including different viewpoints and talents. A skillful facilitator, such as my machine tool industry colleague Jim, acts as both a key contributor and engaged listener. In the best case, the facilitator prepares materials in advance. Such facilitators encourage others to contribute their best ideas while respectfully sidestepping ideas that don't work.

Make sure to showcase and capture ideas as my meeting example did. If you don't use a tool like Miro or Freeform, include a whiteboard recorded by photo. Use low fidelity wireframes for initial brainstorming and later create higher quality wireframes as concepts come into focus. Finally, present and refine higher quality wireframes in stakeholder meetings.

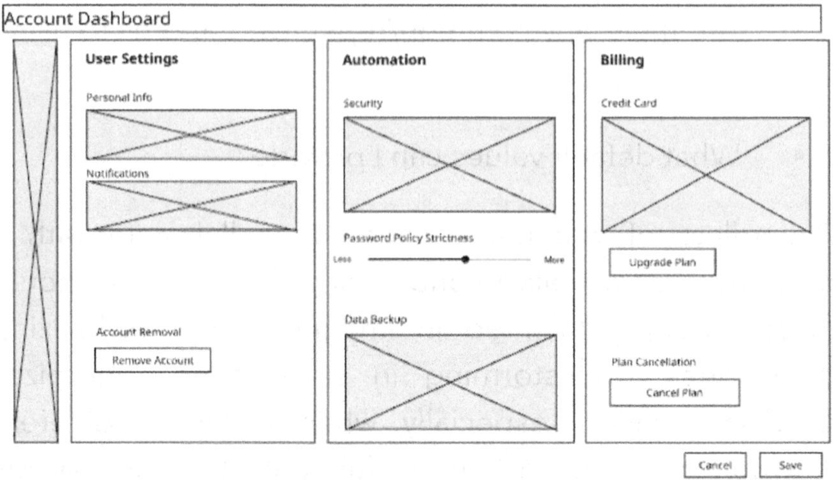

Figure 7: Example low fidelity wireframe — miro.com & author

Whether or not you have direct access, give careful thought to your customer's environment and what it means to work in it. Befriend experts who spend time with your audience such as technical support teams and project engineers. Spend time in the customer environment to gain valuable insights. Is the environment loud? Do people wear protective equipment that might impede UI interaction? How often do they interact with the UI and why? What can you do to draw attention to situations needing attention? How can the UI be unobtrusive when not needing attention?

In a car, drivers focus on the activity of getting to their destination in safety and comfort, not the mechanisms of internal combustion engines or electric motors. Drivers must negotiate obstacles and road conditions that interrupt the monotony between start and destination.

Your ability to embrace the perspective of the "driver" of your UI from outside the mechanism leads to success.

CarPlay Ultra provides the best example of a fully integrated driver perspective available as of this writing. By default, the experience reflects the manufactures style. However, instead of restricting CarPlay to the infotainment system, the experience extends across the entire dashboard. The boundary between the car and the rest of the driver's digital life disappears.

Figure 8: CarPlay Ultra entire dashboard integration - motor1.com

To summarize:

1. Start with thorough knowledge of the customer's goals and a good basic knowledge of your system characteristics. Ask pointed questions that yield detailed requirements. If you can't directly

collaborate, question a proxy such as a local subject matter expert or place yourself physically (or at least mentally) in the target environment. When all else fails, a good imagination coupled with exhaustive system knowledge must suffice.

2. Carefully challenge your assumptions about delivering the status quo, more on this later... Those who take a fresh perspective gain market advantage just like graphical UIs replaced command line interfaces. More elegant interaction with AI will likely replace clumsy prompts.

3. Use wireframes to drive brainstorming and consensus. Generate pixel perfect illustrations only after reaching agreement in principle and only when absolutely necessary. Low fidelity wireframes save time and money while also avoiding a false sense of near-completion for your project.

4. Identify the happy path that meets your customers' goals with the fewest steps and minimal effort. This includes minimizing clicks, gestures, taps, or even voice commands to enter information. A good rule of thumb accomplishes basic tasks with three or fewer clicks, taps or gestures! Also, minimize entry of data by providing default values such as past inputs and favorites.

5. Go to the dark side, *er um*, explore and detail the

exception (failure) cases rather than wait for bug reports, or worse, tech support calls. This clearly beats hopeful chanting of: "Nothing can go wrong. Nothing can go wrong..."

6. Consider extending the UI when needed to perform more complex and less frequent tasks, as opposed to cluttering it with full functionality. Such de-emphasis prevents infrequently used features from distracting attention from features used daily.

7. There is a time and place for a multitude of UI information, but give people what they need, when they need it. Distinguishing frequently used from occasionally used tasks helps solve the dilemma of overly minimalistic versus everything all-at-once.

To further hone your design, let's discover how to delight your audience and avoid overwhelming them as we continue. Exceeding expectations provides a second puzzle piece that we assemble towards you becoming an indispensable agent of a better future. What design challenges might you face and what wisdom from the writings of a human interface expert might inspire you as they did me?

Lightbulb icon from: https://icons8.com/icons/set/idea

CHAPTER TWO

Enhancing Human Strengths and Minimizing Limitations

"Good design is actually a lot harder to notice than poor design, in part because good designs fit our needs so well that the design is invisible"
-- Donald A. Norman, "Father of User Experience".

This longest chapter introduces many core concepts elaborated later. I also honor human interface expert, respected author and mentor Donald A. Norman. Norman is currently Professor Emeritus of The Design Lab at University of California, San Diego and widely considered "Father of User Experience." I sought out his books soon after we started building machine control graphical user interfaces on Windows. His latest book, *Design for a Better World* transcends usability by addressing the role of design in solving larger societal issues. I reference his earlier (and still essential) writings after some background.

Throughout my career, I designed and built tools to simplify complex setup and configuration tasks. These stepwise guides were originally called "wizards". Designers now prefer more descriptive and common terms like "Assistant", "Guide", or even

"Configurator".

In 2005, people consumed the Internet primarily on personal computers. Internet "appliances" arrived two years later with iPhone. Our team stayed entrenched in Microsoft Windows applications. I asked Ken, my decades-long colleague, to collaborate on the connection wizard, calling it my latest "damn fool idealistic crusade" (Star Wars quote, check!). Despite my awkward invitation, Ken agreed. That experience ended up both a delight and portent of future adventures.

As an example, see the below screenshot of the setup wizard from the early days of Windows. (See Figure 9)

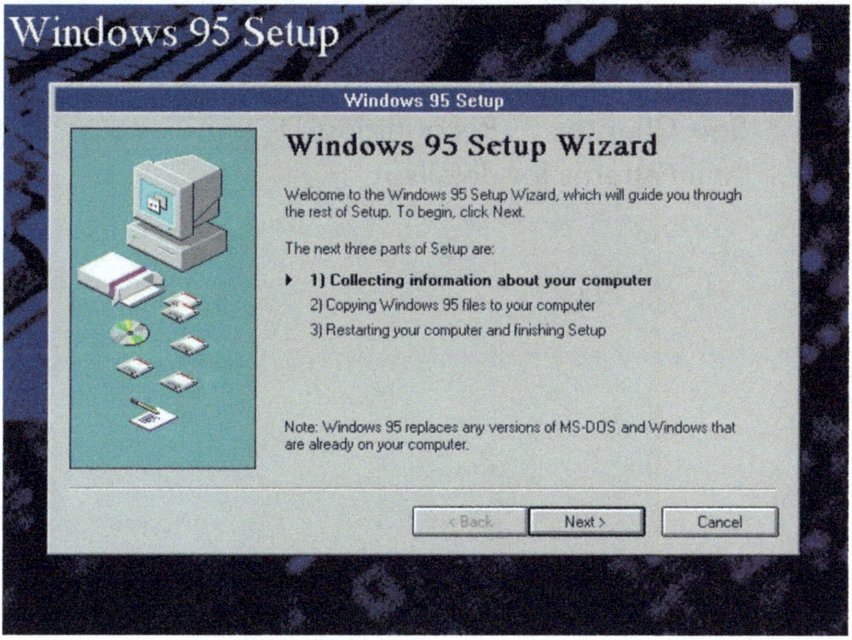

Figure 9: Early Windows Setup Wizard - wikihow.com

Let's explore some tools that succeeded where the Server Editor (previous chapter) failed.

1. Connection Wizard: simplified connecting an off-the-shelf PC to a machine control system. The wizard featured graphics of cable connectors, protocol descriptions and produced a bespoke configuration file. The wizard obsoleted having development engineers spoon feeding files or lines for an XML file. Problem solved!

2. Diagnostics and Configuration Tools: Maintenance and configuration teams must tune and test various material handling subsystems to efficiently move materials in a warehouse. With requested refinements, this tool became a mission-critical maintenance feature. We also established the "Diagnostics Design Pattern" from this experience. See Chapter 3: Recognizing Design Patterns and Anti-Patterns for details.

Before those projects, I started learning design techniques from several of Don Norman's books including *The Design of Everyday Things* and *Emotional Design*. (See Figure 10).

Figure 10: Recent cover of *The Design of Everyday Things* - amazon.com

My summary of key concepts from Don Norman's *The Design of Everyday Things* follows:

1. **Use both knowledge in the world and knowledge in the head.**

2. **Simplify the structure of tasks.**

3. **Make things visible: bridge gulfs between Execution and Evaluation.**

4. **Get the mappings right.**

5. **Exploit the power of constraints.**

6. **Design for error.**

7. **When all else fails, standardize.**

Point 1: Use both knowledge in the world and knowledge in the head

People accomplish many tasks without precise knowledge. A step-wise guide usually makes frequently performed or well understood workflows tedious.

However, creating simple stepwise workflows for occasional and complex tasks builds people's confidence. Configuration tools skillfully turn people's expertise into precise configuration data and delight customers. (Figure 11)

Figure 11: Mini Cooper Car Configurator — minusa.com

A high percentage of people already have a smartphone. Using iOS or Android platform UI conventions (such as appearance and positioning of buttons and toggles) reduces training time and cost. Obviously, an app running on a particular device such as an Apple or Android phone carries certain expectations. Misuse platform conventions at your peril. Instead, reserve cryptic clues or solving puzzles to games where challenge delights your audience.

Infrequently performed actions benefit from stepwise guides built on expertise of their audience.

Point 2: Simplify the structure of tasks

A well organized UI captures the essence of its subject matter. Such elegance makes the solution to a complex problem appear simple and conceals software sophistication behind the scenes. For example: calculate rolling averages and display them as summarized trends and statistics. Paradoxically, an elegant design may seem trivial, but requires great effort to perfect.

Associate menu entries with audience-familiar groupings such as a car's navigation, music, phone and vehicle apps. Knowing how your audience understands their working environment makes order out of chaos.

"Intuitive" derives exclusively from the audience's perspective.

Limit menus to a maximum of 7 items. Organize further into logical subgroups when necessary. Also, remember the relative importance of items in a list. Human memory places greatest importance on the last item in a list followed by the first. Such advantages of name ordering drives actors to seek first or last position in credits.

For automated systems like adaptive cruise control, develop confidence by feeding back AI decisions such as the position of a detected car ahead (Figure 12). For industrial automation, differentiate conveyor stopping to save energy (normal) from a downstream jam (needing manual intervention).

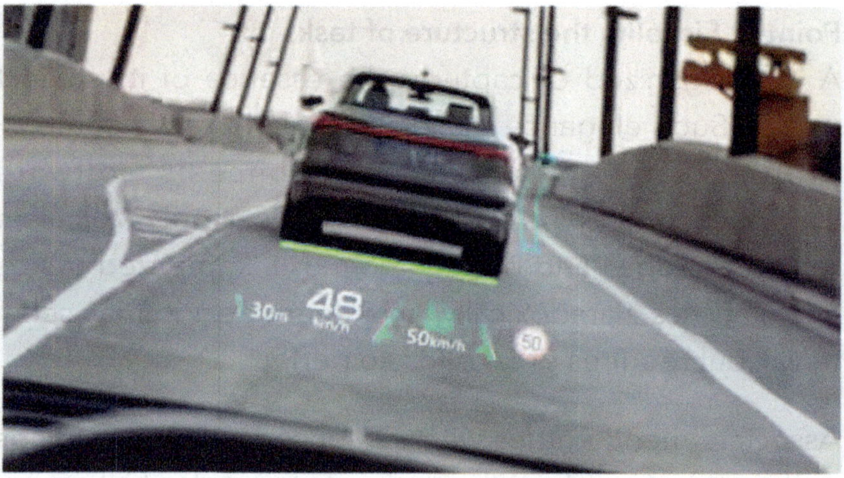

Figure 12: Head Up Display Showing Detected Car — Carwow.com

Only considering requirements from a few unusual or highly specialized customers may also mislead. You must carefully evaluate customer input to generalize a design for a wide range of related, but diverse market segments.

Supporting personalization helps bridge this gap. (More on this in Chapter 7: "Personalizing Appeal in General Purpose Software").

The ideal experience combines hardware and software into an "it just works" ethos of plug and play solutions. If you can accomplish setup or tuning with minimum manual effort and strategically market such advantages, such recognized excellence elevates your product to industry leadership.

Point 3: Make things visible

Make key functionality easy to discover and access while de-emphasizing less used features. People commonly refer to these characteristics as ease of learning and/or ease of use. If you hide common functionality, require too many clicks or make a common use case difficult, frustration and product failure may result.

Instead, draw attention to information with appropriately colored icons. Designers organically choose designs that put status icons on the far left column of tabular data. (See Figure 13). We agreed on leftmost icons for both machine tools and material handling systems alike. Great visual design stands the test of time and spans cultures.

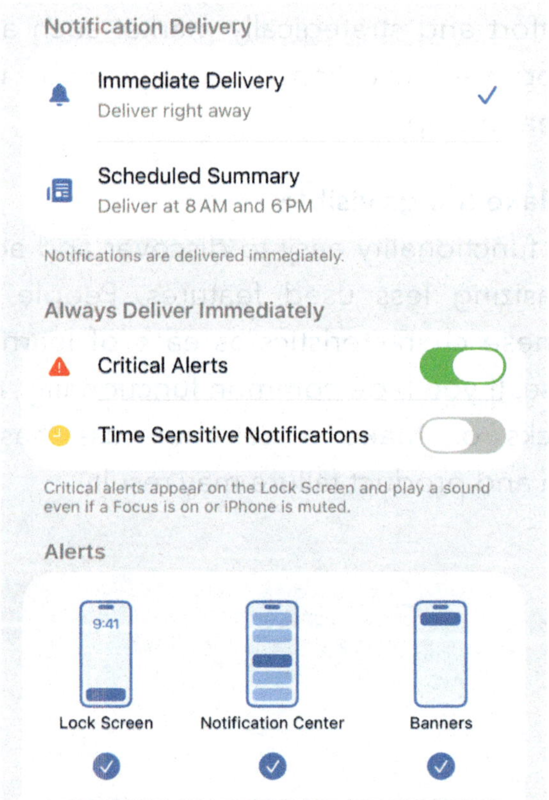

Figure 13: Effective icon use — iPhone Weather Settings

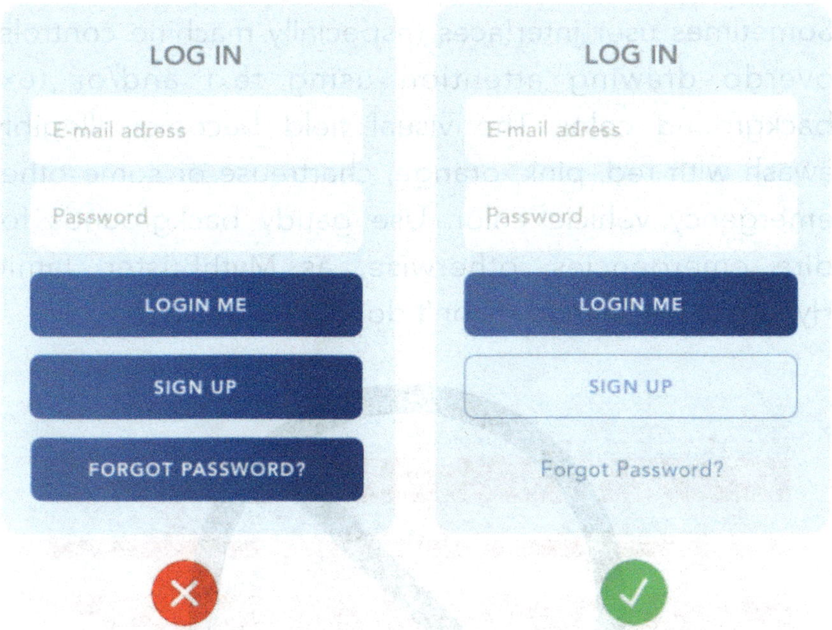

Figure 14: De-emphasizing less typically used features (right) — careerfoundry.com

Draw focus where and when needed using icons and graphics. Avoid gaudy backgrounds.

Sometimes user interfaces (especially machine controls) overdo drawing attention using text and/or text background color. The visual field becomes illegibly awash with red, pink, orange, chartreuse or some other emergency vehicle color. Use gaudy backgrounds for dire emergencies, otherwise, as MythBuster Jamie Hyneman disclaimed: "Don't do it!" (Figure 15).

Figure 15: Avoid overwhelming illegible backgrounds — sisautomations.com and author

Instead, use colors carefully and validate appropriate contrast with the WebAIM contrast checker.

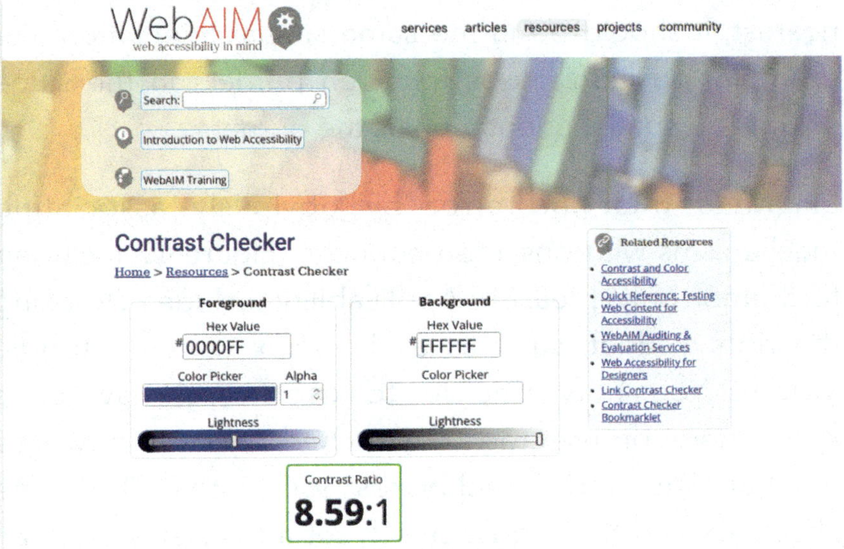

Figure 16: WebAIM Contrast Checker — https://webaim.org/resources/contrastchecker/)

Our experience identified a much bigger problem with alarms than just a sea of red. People need to view only pertinent alarms to minimize downtime, not treated to dozens of side effects. Jesper, a machine control SME, set establishing root cause as the primary goal. His criteria identified proximate causes and graded them. A filter selected which alarms to show and hide. Operators could show or hide surrounding alarms as needed, thus realizing Jesper's vision.

Point 4: Get the mappings right

People click/tap/press the button (or turn the knob) they want, more reliably, when you position it as similarly to the device as possible. Preferably the button is on the device. If not located on the device, it should lay out nearest to the physical device that it controls. If not nearest, it should be in the same spatial orientation. For example, position knobs for oven burners in the same layout and spatial orientation as the burners.

Eliminate trial and error by obviously associating mechanisms with onscreen controls. (Figure 17 included for humor only, please!). Applicability includes choosing the right widget, such as a checkbox versus a toggle switch. You may decide to use toggle switches everywhere or use toggle switches only for powered mechanisms and checkboxes for states. Both the checkbox and the toggle switch set a true/false boolean variable in code, but adhering to rigorous guidelines makes all the difference!

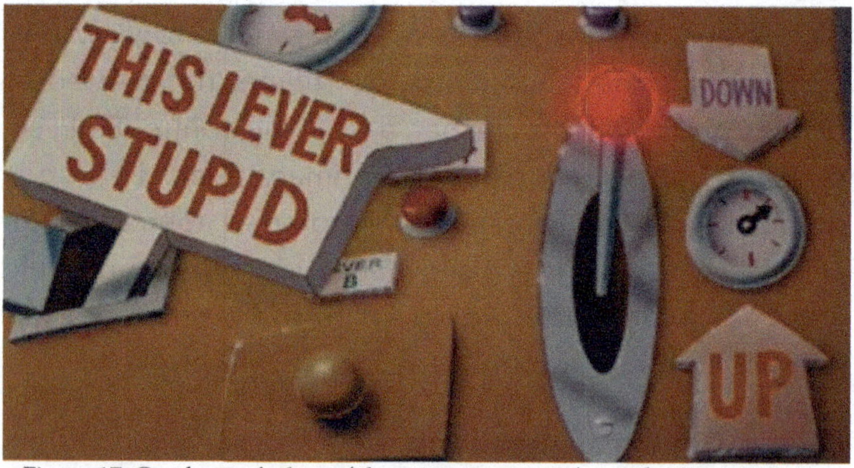

Figure 17: Good use of a lever (obnoxiousness aside) — Who Framed Roger Rabbit and tvtropes.org

If the value cannot be changed where it is being displayed, choose from the following possibly applicable standard labels.

1. Yes/No: for most mobile UI settings.

2. On/Off: for powered mechanisms.

Never display a checkbox or toggle switch for a Boolean true/false state when people cannot change the value. Display "Yes" or "No" text instead.

Don't use "True" / "False". Only developers think this way. People may figure this out, but most people find such "Boolean" labels unnatural.

Be aware that toggles may confound clarity. Are standalone icons indicating action or state? For example a turn by turn speaker icon could indicate it will go off if you touch it or the current state is off. Whatever your choice to show state, apply the rule consistently!

To summarize: display modifiable widgets when editable and text (or graphics) otherwise.

Point 5: Exploit the power of constraints

A two or three position switch found on many control panels demonstrates using constraints in your designs. Another includes a drop down list with an appropriate number of responses (best case, between three and seven entries). Both cases limit choices to only the appropriate ones. Lists and autocompletes function far better than requiring typing when your software may interpret a person's typed text multiple wrong ways. (Figure 18)

Figure 18: Autocomplete dialog box — telerik.com

Alphonse Chapanis (March 17, 1917 - October 4, 2002) pioneered industrial design and human factors in the USA. He published *Applied Experimental Psychology: Human Factors in Engineering Design* in 1949 (Source: psychologicalscience.org).

Chapanis notably discovered an early UX design error back in the 1940s. B-17 pilots would mistake similar looking flap controls for landing gear and vice versa, resulting in severe damage. Would you confuse the pictured switches? (Figure 19)

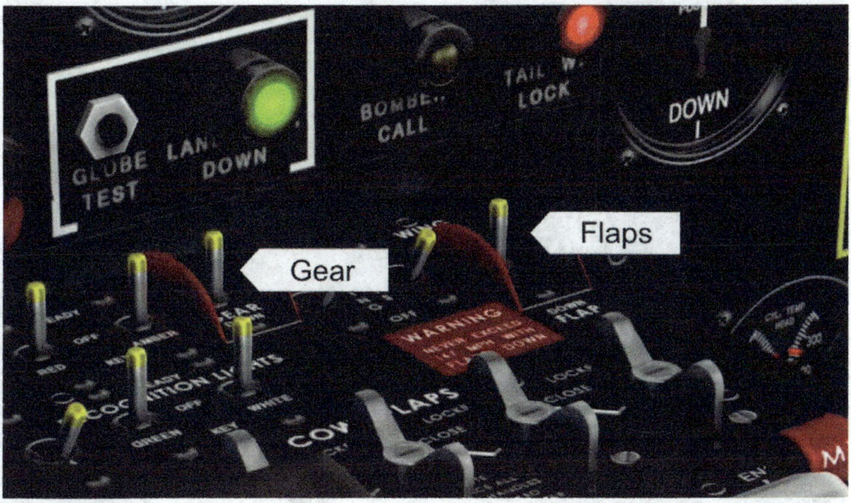

Figure 19: B-17 Confused Gear and Flap Controls with added red channels — uxmag.com

Chapanis proposed shape-coding aircraft controls to reduce errors and allow operation by touch alone. These standards remain in effect today now expanded to at least 11 different aircraft controls.

Orion spacecraft switches for the Artemis program include guards to prevent accidental activation by astronauts floating in weightlessness. Space Shuttle cockpits featured the same kind of switches. (Figure 20)

Figure 20: Orion Spacecraft Simulator — asc-csa.gc.ca

People often accidentally activate touch screen controls. For risky touch controls, include interlocks such as long presses and/or confirmation boxes. In cars, people prefer more tactile physical switches and knobs they can set while keeping eyes on the road. I've also turned on the defroster by using my voice and the physical voice command switch. Fantastic!

Never ever use anything other than hard-wired physical buttons for safety buttons such as Emergency Stop.

Government regulations now mandate hard-wired emergency stop buttons and cords. Regardless, reading this book means you want to do the right thing.

Notice also the special purpose switches distinguished by yellow and black striping (in the above figure)

Carefully differentiate confused actions with the appropriate widget in the appropriate orientation (up/down, left/right, etc).

Constraints apply similarly to supporting cultural conventions such as reading left to right, (or in some cultures right to left). Unless you want to be unintentionally funny or profane, beware! Finger gestures vary in different cultures. Be especially wary to avoid icons featuring possibly misinterpreted gestures (especially those involving fingers or the giving thereof 😬).

Some naming trickery accomplishes both an alphabetical and importance ordered list, but unless your framework automatically orders these lists, alphabetical lists fall apart when localizing your product.

Point 6: Design for error.

Common sense recognizes that people make mistakes and processes fail for multiple reasons. Performing multiple calibration passes and allowing result adjustments improve people's confidence in precisely setting calibration. Provide undo where possible, but alternatively require deliberation to avoid accidents. People must understand the consequences of an action when non-obvious and especially when under duress. (Figure 21)

Figure 21: Error dialogs with (some) remedies — dribbble.com/shots/23570984

Provide easy error recovery (such as undo) whenever possible. Otherwise, make unrecoverable actions difficult, so people think twice before proceeding.

Confirmation dialogs may include interlocks to further reduce the chance for error. One technique prevents continuing until checking a checkbox. Such features stand in for undo when situations don't allow reasonable reversal. At the very least, an accidental shut down equates to lost productivity. More extreme cases lead to an unrecoverable erased hard drive.

Point 7: When all else fails, standardize.

This point requires both serious consideration and parodies itself. Donald Norman calls out both what to do in the absence of an intuitive solution and tendency of developers to always believe in a "better way". Fix what you find broken, make groundbreaking improvement, or otherwise leave it alone. You know the cliche!

Frustration and mistakes result from failure to follow certain standards especially as they apply to a platform such as Android or iOS (Figure 22). One example: consistent positioning of "Cancel" and "Save" buttons. Following an inadequate or misfitting standard is a different matter!

Figure 22: Apple and Android Design Guidelines — developer.apple.com and m3.material.io

Use familiar techniques and terminology found on widely used devices (such as smartphones).

From a different perspective: establish and follow standards within your product. See Chapter 3 and Chapter 6 for more details. Sometimes great design fails to make operation obvious, but if people learn the position of critical buttons such as "Cancel" and "Save", following the convention avoids errors and frustration.

Be alert for subject matter experts designing separate UIs

for each subsystem in a system. Recognizing consistent patterns requires acquiring some in-depth system knowledge. You don't have to understand all the underlying automation, but you must understand real world basics of operating your system. If you fail to standardize with a hard won grasp of fundamentals, the result will be frustration, long learning curves and more mistakes.

This chapter introduced many important concepts without full details. Don't worry, I elaborate all these points and more throughout the remainder of this book. Expect details of "how" and "why" such as: How do you avoid civil liabilities from consistency mistakes and misguided customer trickery?

CHAPTER THREE

Recognizing Design Patterns, Anti-Patterns and Dark Patterns

"Order and simplification are the first steps toward mastery of a subject"
-- Thomas Mann, Novelist, Short Story Writer and Nobel Prize Winner

You can save your employer and your customers considerable time and money by recognizing repeating patterns in your application. Your audience experiences a streamlined and standardized experience. I established the use of design patterns in my experience below:

In 2012, I joined a leader in the material handling industry (see Figure 23). My good fortune was needing a job change and landing an interview, thanks to my colleague Ken who joined the company before me. The company found itself on a venture capital and growth driven hiring spree. The existing team already mastered fundamentals necessary for industry leading automation and software infrastructure. Ken claimed I could improve their UI.

Figure 23: Material Handling System — mmh.com

Another member of the team designed bottom up and I offered a people-centric alternative with the comfortable familiarity of Apple's iOS. The team ultimately decided to go with my design, but untangling the daunting knot of UI inconsistencies was a constant of my remaining career.

The leader of the automation team dubbed my role as "Consistency SME". To unify ideas of other subject matter experts, I applied "Design Patterns" from the pioneering works of the "Gang of Four" (aka GOF: Gamma, Helm, Johnson and Vlissides). (See Figure 24). The GOF considers a design pattern as a detailed set of instructions for building reusable software components. Design patterns recognize repeating features in complex software. This technique allows engineers to reuse and document "best practices" for solving common problems. Instead of using design patterns just to implement infrastructure, we applied them systematically both in implementation and consistent "user-centered" design. An explanation of all this to follow...

Figure 24: *Design Patterns* Book — amazon.com

Engineering leadership had painstakingly built a team of subject matter experts laser-focused on various crucial parts of the conveyor system. Each SME operated almost autonomously, thus sacrificing similarities of subsystems. (Note: Each subsystem consists mostly of similar conveyor pieces). SMEs also designed their own user interfaces from a hardware perspective (bottom up). Therefore, each subsystem had an entirely different user experience.

Independent SME designers can lose customer perspective resulting in "designed by engineers for engineers". In stark contrast, customers hold one operator and maintenance team responsible for all conveyor subsystems. Although each subsystem features unique inner workings, customers benefit from consistent operation systemwide.

We talked about Wireframes in Chapter 1: Getting On Track. Design Patterns help translate these wireframes into a coherent design by elaborating experience details insufficiently described by wireframes.

My colleague Will in the project engineering team relentlessly sought to supplement our narrow property style editor view with a big picture view of settings. We discovered narrow views appealed to maintenance teams and wide views to project engineering teams. We imagined a better redesign featuring a spreadsheet style editor. A spreadsheet layout allows each team the option of comparing and editing settings across multiple related assemblies within each subsystem. Using the Mac's Finder as an example we see properties of a single file versus multiple files. (See Figure 25)

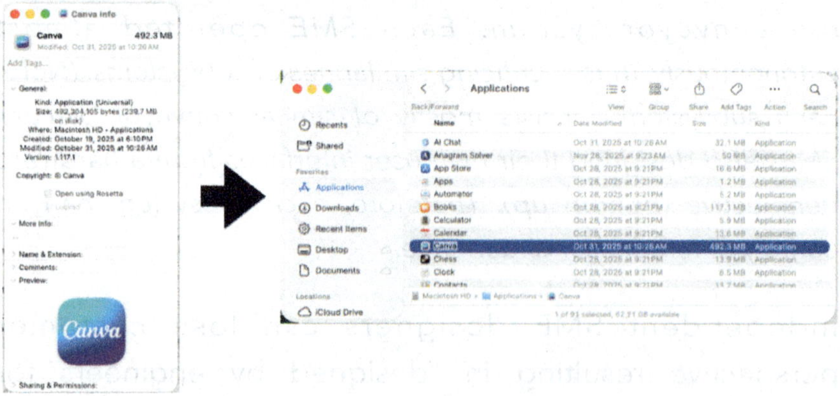

Figure 25: Property Editor vs Spreadsheet Editor - author

Material Handling systems operate much like a train yard, routing trains to many potential destinations. Warehouses may provide access through upwards of 100 doors. Each incoming package possibly routes to any destination door.

Using analogies alters your perspective and encourages innovative design solutions.

When listed together, applying the corresponding settings across multiple table rows both speeds data entry and helps spotting outliers (possible mistakes). Furthermore, selecting the same settings across multiple tracks (or in this case, conveyor lanes) allows spreadsheet style fills. A fill acts as a massive automated paste, thus eliminating tedious data entry. For us, such characteristics required multi-level editing patterns to support complex hierarchies.

Calibration illustrates another complex and seldom performed workflow. Calibration steps may include moving a mechanism into position, placing objects in the workspace and/or performing distance or speed measurements (preferably automatically). During calibration, progress indicators confirm completion of each subtask while setting expectations for remaining effort. Results appear at the end of the calibration for visual verification and/or adjustment by the maintenance team.

Consistent implementation critically requires detailed design patterns, but some differences may lead to simply skipping unnecessary steps. You can think of complex patterns gracefully degrading into simpler ones as needed. Examples include:

Must a maintenance engineer enter and exit a "Diagnostic Mode"? Production may behave considerably differently than maintenance processes. Diagnostics might also degrade system performance requiring the UI to clearly indicate the current state of "Diagnostic Mode". Failure to call out significant mode changes confuse the operations team, especially when maintenance teams accidentally leave the system in diagnostic mode. Our UI encouraged diagnostic deactivation on exit and prominently displayed diagnostic mode icons to reduce support calls (Figure 26).

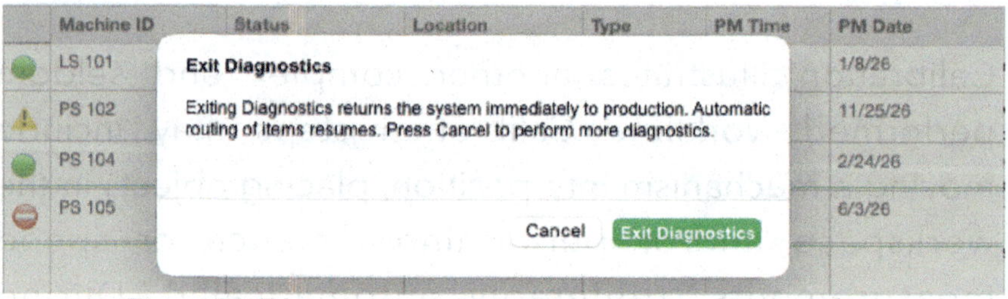

Figure 26: Encouraging return to normal operation - author

Other important questions included:

- Can project engineers modify calibration results?
- When should results be modified? If multiple passes determine an average, how does the

project engineer discard or adjust for an errant calibration pass?

- How do you handle restoring functionality with a required mechanism reset when calibration ends?
- What if there are only some calibration or diagnostic procedures that require mechanism resets?

A good example of a calibration tool appears below. (Figure 27) I made a few tweaks to an otherwise solid design, such as removing borders of non-editable fields and using icons rather than colored backgrounds. The changes follow standards set in Chapter 2: "Enhancing Human Strengths and Minimizing Limitations."

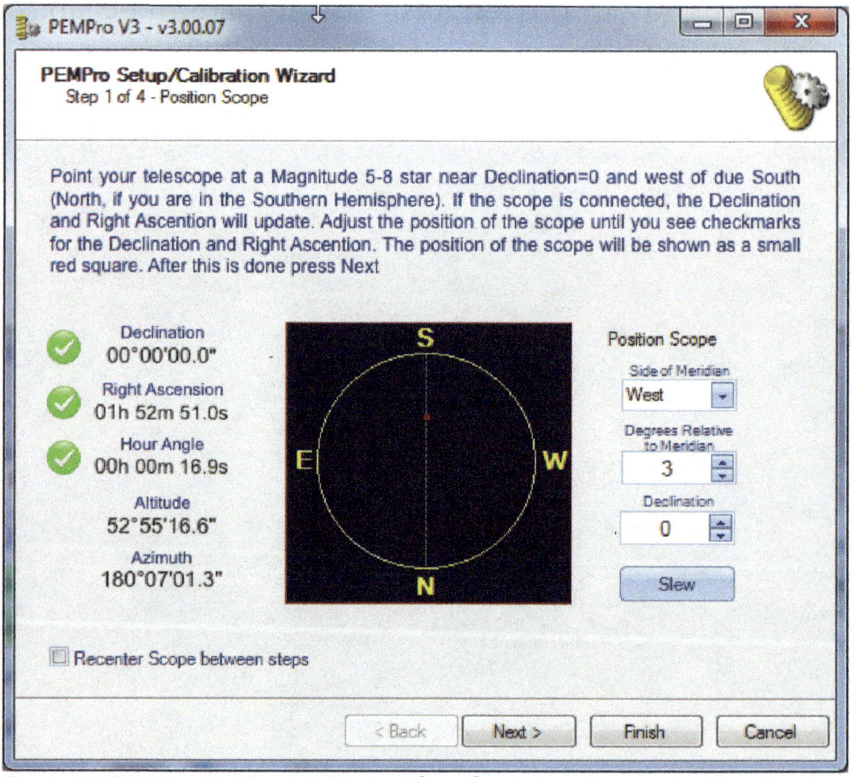

Figure 27:PEMPro Calibration Tool with author's tweaks — siriusimaging.com

Solving each of these problems with precise design patterns guides colleagues toward consistently correct results and eliminates rework. Not all workflows (such as calibrations) require all steps. That's OK if you provide guidance when needed. Don't forget to provide graceful recovery from mistakes and failed test runs. Nowhere are your efforts more appreciated than in anticipating and handling expected errors and avoiding unexpected errors.

Expected errors result from common mistakes or device failures. Unexpected errors result from software or hardware faults.

Expected errors:

- Media removal (ejecting a flash drive)
- Network failures
- Out of Wi-Fi or cellular range
- File not found (invalid search specification)
- Low storage
- Jams
- Emergency Stop

Unexpected errors:

- Storage read or write failures (failing device)
- Database errors
- Memory checksums
- Software exception handling (not resulting in an expected error)

Expected errors require clear explanation of remedy (more on this later). Unexpected errors need reporting so that tech support and ultimately developers successfully address them. Ideally there are no unexpected errors!

Alternatively, If you design for an existing platform, certain patterns and style guidelines already exist. Such patterns help translate your wireframes into finished software without having a design team generate pixel perfect representations for each screen layout.

Other examples of guidelines from design patterns:

Prevent buttons from disappearing and reappearing unexpectedly such as for a loss of connection or change of state. Neither allow such a result to override changes at a different operator station. The nightmare scenario results when an operator wants to start a machine and the function of the button changes between the time the operator goes to press the button and the time it actually takes effect.

If the button function changes from start to stop mid press (not recommended!) a costly shutdown might result. This yields a recipe at least for unhappiness and potential danger. Safety conditions require either a warning or a confirmation dialog box. Also confirm operations for buttons resulting in downtime and requiring significant time to reverse.

As previously stated, consistently lay out Cancel and Save buttons for dialog boxes (and all common buttons). Hopefully you consider this obvious, but when unenforced and with larger/inexperienced teams, inconsistent mayhem may ensue. The button shown last in order holds greater significance, especially when placed on the right side of a dialog box. Therefore, show the safe action (Save) on the far right and emphasize it according to your style guidelines. Feature unsafe or dismissing actions less prominently. For those reasons, Cancel and Save appear in the order and emphasis shown. (See Figure 28).

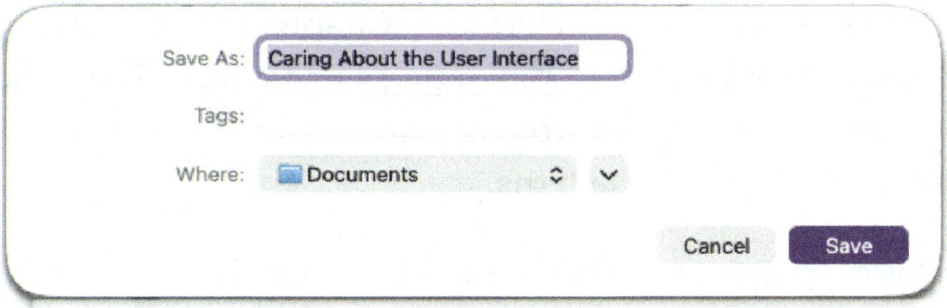

Figure 28: macOS Tahoe Save Dialog showing button location and emphasis - author

Document consistent order and placement of related buttons. (Preferably enforce with frameworks (Chapter 6))

Other repeating patterns include different subsystems in a facility that need to be started and stopped, enabled and disabled. Failure to standardize design elements and workflows means your customer experiences a hodgepodge contributed by individuals working in their own respective silos.

Sometimes different teams start with different solutions to the same problem and you must determine the best practice solution and standardize on it.

Standardize before going to market **and** after discovering improvements.

The evil twins of design patterns are so called "anti-patterns" and "dark patterns".

Anti patterns arise from failures to follow standards. We talked about these failures in Chapter 2: "Enhancing Human Strengths and Minimizing Limitations." They also result from failure to recognize and consistently implement design patterns.

Design pattern standards should also establish a tone for your dialog boxes (and confirmation dialogs). We avoided the age old throw away verbiage of "Are you sure?" Or "Do you want to continue? Instead, I heartily recommend emphasizing the consequences of any significant action.

I construct the hopefully hilarious example of many anti-patterns below. Don't let this happen to you:

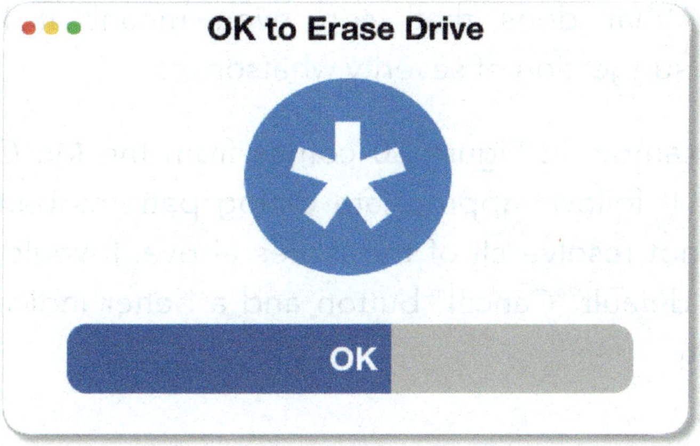

Figure 29: What's wrong with this picture? - Author

How many of the following blunders did you find in the intentionally wrong dialog box example (Figure 29). Here are a few:

1. Desperately needs a prominent "Cancel" button as the default action!

2. The "OK" button should be re-titled "Erase Drive". Some people go directly for the button without reading or understanding the dialog.

3. It should not automatically take action on a timeout! However, designers underutilize timeouts for situations that might greatly benefit.

4. Window resize or close buttons should not be at the top.

5. No description of the action consequences included.

6. What does that icon even mean? It has no suggestion of severity whatsoever.

The example in Figure 30 comes from the MacOS Disk Utility. It follows appropriate dialog patterns better, but does not resolve all of the issues above. I would like to see a default "Cancel" button and a better indication of severity.

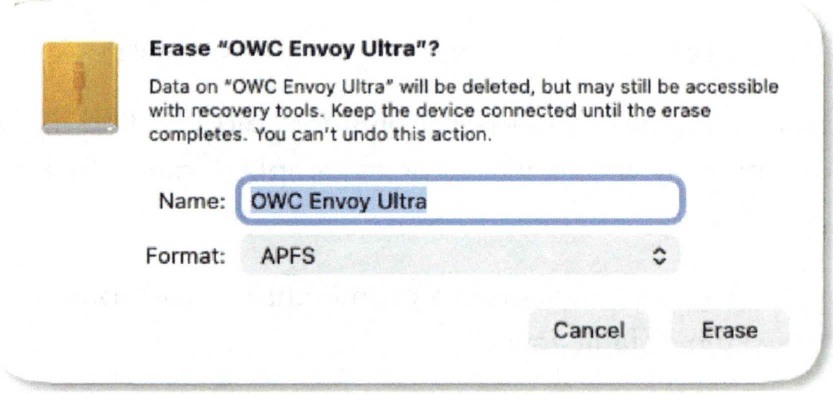

Figure 30: Mac Disk Utility Dialog — Apple, Inc.

Dark patterns arise when companies try to self-servingly manipulate behaviors. For example, sometimes UIs make it difficult to cancel subscriptions or opt out of marketing promotions. The prevalence of dark patterns in software provides a good measure of a company. Some industry titans like AOL took this path when their subscription based business models collapsed, but sadly, current examples of this practice also exist.

If a business can't be monetized in some way, it typically fails. However, there is a line between "We're sorry to see you go" when you unsubscribe from emails and dirty tricks. As one example of dirty tricks, make people remember the email address they used to subscribe. As another example, require lots of scrolling or investigation to figure out how to unsubscribe.

Many examples exist of the so called "Roach Motel" pattern. Simply put, it is easy to get into, but nearly impossible to exit.

Cited examples include roadblocks to:

- Checkout without subscribing. You have to bypass numerous ways to checkout with a subscription before being allowed to checkout without.

- Find out how to cancel your subscription. Websites provide no search results for how to cancel.

- Successfully completing an unsubscribe. You have to click and scroll through multiple pages of discouragement such as discounts. You could locate another dialog box to uncheck automatic renewal only if persistent enough.

You might think you successfully cancelled only to discover later you hadn't. The insidious nature of these patterns prey on less tech savvy and less attentive. They establish customer animosity rather than loyalty.

Bad actors use "algorithms" that all too easily manipulate economic and political behavior causing suspicion and mistrust. Cambridge Analytica used Facebook profiles to build "psychographic" profiles of users to target American voters with targeted political ads that helped influence the outcome of the 2016 political election (sources: The New York Times, Amnesty International, et al). This example illustrates the potential harm of products using algorithms to target content.

Even more troubling, companies may foster addiction by pandering to people's biases. According to a March 26, 2026 New York Times article (by Cecilia Kang, et al), a court in California ruled against Meta and YouTube citing design features that "led to her mental distress". The lawsuit chose to cite the entirety of Meta, but not YouTube's parent Google. Such litigation will probably rise to the high courts as a wide variety of similar suits confront the court system from individuals, school districts and class actions.

From a designer perspective, the California lawsuit pointed out infinite scroll and algorithmic recommendations as examples. Please note: I believe the intent and application of these features are distinct from the features themselves. I've advocated for patterns similar to infinite scroll for large tables. For news and social media feeds, soft limits and "up-to-date" indications might discourage some endless "doom scrolling."

Thankfully, over time, people learn to better handle dark patterns, but caring designers need to consider these examples of what to avoid and be aware of the consequences of our decisions. Threat of litigation further incentivizes your business to avoid dark patterns. I found a lot more details in articles like this one on the eleken.co on dark patterns examples.

Documented design patterns help promote consistency, but what sources of inspiration might help you create robust and well accepted patterns?

CHAPTER FOUR

Persisting Vision and Honoring Ideals

"Knowing the rules and when to break them"
– Candice Olson, designer, HGTV host , et al.

Documented patterns help establish and share the essence of good design. There are excellent examples of UI design everywhere. Powerhouses like Apple and Google painstakingly research and continually refine their patterns. Such patterns go by "Material Design" (Google) and "Human Interface Guidelines" (Apple). Study these guidelines and apply them where appropriate. Notice their commonality. Avoid awkward reinventions. Also, immerse yourself in the mastery of acclaimed apps such as Apple, Google and Webby award winners (webbyawards.com). Come away with questions like:

How do people know how to accomplish their objectives? Must they reference documentation or instead accomplish their work unguided? Better yet, do people already know how to work intuitively based on knowing standard patterns and/or follow hints made to encourage discoverability? Do you see opportunities to reorganize, streamline or reduce annoyances?

User Interface should drive APIs (and object models), not the other way around.

I worked in multiple industries where we recognized utility in both overview graphic and table display patterns. As we considered reinventing our next generation UI, the product offering team wanted to express the entire UI in 2D or 3D graphics form. (Figure: 31)

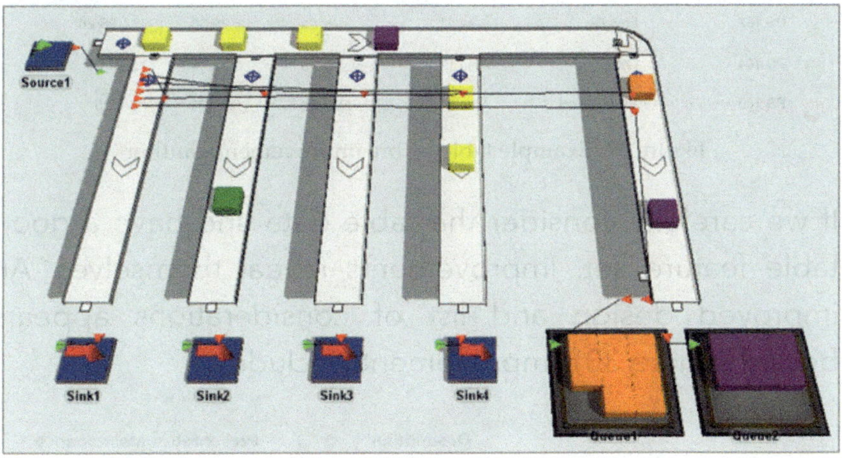

Figure 31: Conveyor Graphical View — flexsim.com

While we found graphical 2D and 3D displays best at locating and resolving simple operating faults, tables remained vital. A well worn cliche states: "A picture is worth a thousand words". However, Bruce Tognazzini (Ask Tog) emphasized "a word is worth a thousand pictures" as demonstrated in Figures 32, 33 and 34.

Recurring design patterns don't happen by accident. With each redesign, designers repeatedly discover the utility of data tables as effective patterns for rendering densely packed data. Correctly implementing such patterns requires studious attention to a multitude of details. Together, these details make the difference between burdening or empowering your audience.

Let's explore an example! What pattern details apply below? (Figure 32)

	Machine ID	Status	Location	Type	PM Time	PM Date
●	LS 101	OK	Area 101	Rapid Sort	4:12:25	1/8/26
⚠	PS 102	Low Oil	Area 51	Vectra 245	5:29:38	11/25/26
●	PS 104	OK	Area 12	Vectra 391	12:37:02	2/24/26
⛔	PS 105	Stop Pulled	Area 9	Vectra 823	0:00:00	6/3/26

Figure 32: Example table before improvement - author

If we carefully consider the table data and have a good table feature set, improvements reveal themselves. An improved design and list of considerations appears below. (Figure 33). Improvements include:

		Description		Preventative Maintenance	
Status	Machine ID	Sorter Type	Location	Date	Time
● OK	LS 101	Pop Up Wheel	Mezzanine	Thursday	4:12:25
⚠ Low Oil	PS 102	Cross Belt	Jackpot	11/25/26	5:29:38
● OK	PS 104	Pop Up Wheel	Shipping	2/24/26	12:37:02
⛔ Stop Pulled	PS 105	Cross Belt	Receiving	6/3/26	0:00:00

Figure 33: Example table with improvements - author

1. Leftmost column icons draw attention to issues needing urgent attention. Differentiate icons by

color, shape, and symbol. Minimize the total list of status variations (preferably 7 icons or less). Sometimes a picture is worth a thousand words and sometimes a word is worth a thousand pictures. Notice the happy combination of both shown leftmost.

2. Left alignment of textual information such as names, locations, types

3. Substitute a recognizable name like "pop up sorter" rather than branding text unless dealing with inventory labeled primarily with branding. Even then, providing a generic type helps with finding and grouping similar items. Make similar more descriptive changes for location. Humorous, but valid: Sorting facilities often use the whimsical term "Jackpot" to describe backup destinations for items failing to route where intended. Allowing name customization provides another alternative we discuss in Chapter 7: "Personalizing Appeal in General Purpose Software."

4. Center alignment for pictures and icons when they appear without text.

5. Right alignment for numbers, times and dates along with their column titles.

6. Using a day of the week (such as "Thursday)" rather than a date (December 4, 2025). For dates like

today or this week, using the day of the week or "Today" frees people from mentally calculating the proximity of a near term date (or time). Grouping Date and Time logically. Note: Date and time formats vary due to localization considerations (also in Chapter 7)

7. Alternating row colors make the rows easier to read across.

We considered many more issues including:

- Should tables offer touch compatibility and thus larger cell padding (vertically at least). Maintenance teams generally prefer more information dense displays. Consider making features like vertical sizing and padding configurable. Better yet, use responsive design where possible to adapt automatically.

- What priority applies to each column from left to right? We got feedback of appropriate priorities and accordingly standardized column order. People could also save column reordering and later reset to our defaults.

- Does your table widget support grouping of columns? Column Groups create more easily understood data "chunks." (Like "Description" and "Preventative Maintenance" column groups).

- Should you support horizontal scroll and if so, should you freeze one or more leftmost columns? Data loses context when status and names scroll off.

- How useful are resizable columns? More useful in situations of high information density but set sizes to typically adequate defaults.

- How can you tell if text overflows? Append an ellipsis (...): for text on the right (for left to right languages). Numbers and dates should favor the most significant digits.

- How do you handle data much too large for a column? Split views solve this problem and optionally allow people to resize the split as needed. Also, use split views for child tables.

- How do you call out information hidden in large lists? Use summaries above data tables to emphasize exceptional conditions and allow filtering to quickly view them. (See below)

Errors 2 Warnings 1 OK 15

		Description		Preventative Maintenance	
Status	**Machine ID**	**Sorter Type**	**Location**	**Date**	**Time**
OK	LS 101	Pop Up Wheel	Mezzanine	Thursday	4:12:25
Low Oil	PS 102	Cross Belt	Jackpot	11/25/26	5:29:38
OK	PS 104	Pop Up Wheel	Shipping	2/24/26	12:37:02
Stop Pulled	PS 105	Cross Belt	Receiving	6/3/26	0:00:00

Figure 34: Example table with summary and scroll enhancements -author

For many environments, the best user interface usually blends in and only commands attention when urgent conditions arise (hopefully before urgency becomes emergency). Especially in industrial environments, people multi-task and don't want to be bothered unless situations require their attention.

Similarly for drivers, appropriately differentiate urgency of an imminent crash versus upcoming scheduled maintenance. Don't get me started about the tragically useless "Check Engine" light. Either send a message to a technician or display actionable information, please!

Here are some rules of thumb for concise display of essential data:

- Provide vital information without distractions. Icons effectively draw attention as these examples demonstrate. Figure 35 shows the difference between distraction (crossed out) and at a glance reassurance.

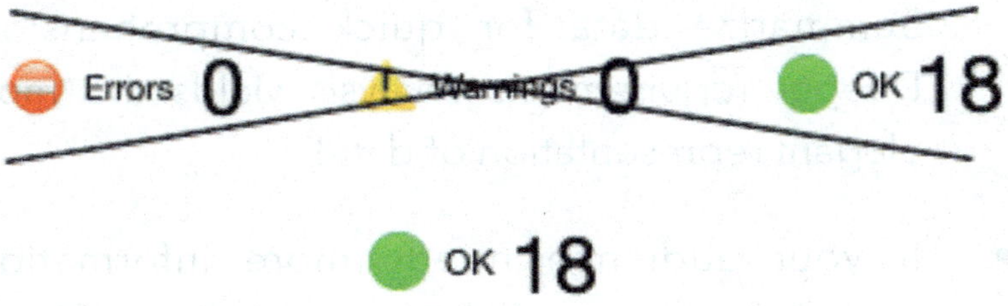

Figure 35: Distraction versus reassurance

Emphasize and prioritize vital information at a glance.

- If a fault condition occurs, draw the responsible person's attention. Often, different people fix a stall or jam condition versus an equipment failure. In many facilities, seconds of downtime are lost productivity which equates to higher cost and weaker competitive advantage.

Include location, cause **and remedy** with error conditions. Actionable fault data drastically reduces downtime.

- Prominently feature data your audience really needs. Would people rather route through thousands of events to determine trends or for your software to proactively analyze and summarize the data?

Summarize data for quick comprehension. Deeper requirements analysis yields the most elegant representation of data!

- If your audience needs more information, provide tools to drill into specifics. Efficient resolution of malfunctions potentially saves lives and, at the very least, profits.

The techniques in this chapter lead to a cornerstone of UI design.

Never throw data together. Make sense of what you choose to show and how you show it!

Nothing brings customer delight more than software that elegantly and insightfully achieves their goals.

We've explored optimizing design patterns, but what secret weapon first draws people instinctively to your product or service?

CHAPTER FIVE

Appealing with Beauty and Elegance

"People tend to believe things that look better will work better"
-- Nielsen Norman Group, UX consultancy founded by Don Norman and Jacob Nielsen

The quote for this chapter comes from an excellent article linked on nngroup.com.

Achieving real beauty requires designer attention to more than just aesthetics.

Starting in 2016, our group built a powerful web-based application UI that lead the industry in many ways. Only a few years later its "dated" appearance (and corporate politics) drove its replacement. As we worked towards a more modern and beautiful UI, we also evaluated improvements to workflow and customization.

If your UI wows with visual appeal, people notice and want to learn more. Apple debuted flat design on the iPhone which seemingly made previous 3D-like (skeuomorphic) designs out of fashion (see Figure 36).

People prefer a "clean look" which results from an attractive layout, easy-to-read fonts, and efficient use of space. Straightforward, minimal text contributes to perceived elegance.

Figure 36: Fully skeuomorphic iOS 6 on left and (partially) flat iOS7 on right - designmodo.com

Our corporate design team helped establish initial design patterns and consistent layout rules. However, a tug-of-war ensued between designer minimalism and our power user group. Designers favored ample padding for elegance, visual simplicity and touch friendliness. Power users craved densely packed screens. Conflicting impulses like these may also call for customization, but more on that later.

Responsive design and media features allow designs to automatically adjust between mouse (@media (pointer: fine)) and touch (@media (pointer:coarse)) precision.

Responsive design beautifully adapts composed screen elements for portrait, layout or even folded/unfolded orientations.

You've probably bought products based on aesthetically pleasing marketing images. I have. Many graphics applications sell themselves by making aspiring artists (like me) think we can magically create the art featured in marketing materials. Gorgeous outcomes still require talent to realize best results.

Let's examine great examples from past and present.

Original MacPaint appears somewhat dated today (see Figure 37), but stunned many on launch. The graphical UI demonstrated first glance superiority over character driven, low resolution competitors.

Figure 37: MacPaint Original Marketing Image - en.wikipedia.org

Certainly use of color, higher resolution and animation improve the appeal of modern software. Your audience expects precisely rendered graphic designs, especially those seeking a luxury experience.

For comparison, here is a recent marketing image for Pixelmator Pro (Figure 38) showing how far modern computers have evolved since the original Mac's low resolution monochrome display.

Figure 38: Pixelmator Pro Marketing Image - apple.com/pixelmator-pro

Beauty also naturally arises from simplicity and elegance. Don't consider your product complete after it gloms on a smorgasbord of features. Instead measure completion against delivering essentials and deemphasizing or removing non essentials. Frustration results from common tasks requiring many clicks (typically more than 3) or when people sink under the overwhelming weight of needless complication.

On the occasion of Apple's 50th birthday, Tim Cook talked about his realization as he looked over years of Apple prototypes and realized the paradox of simplicity. Cook said: "It's hard doing simple. It's easy doing complex".

People perceive the most elegant solution as also the most beautiful.

Our audience, especially "power users", wanted context menus. After running our UI on the engineering loop I wholeheartedly agreed. People instinctively point to on-screen objects and enjoy revealing a short list of relevant actions (Figure 39). Trackpad gestures and long press on touch devices also support handy context menus.

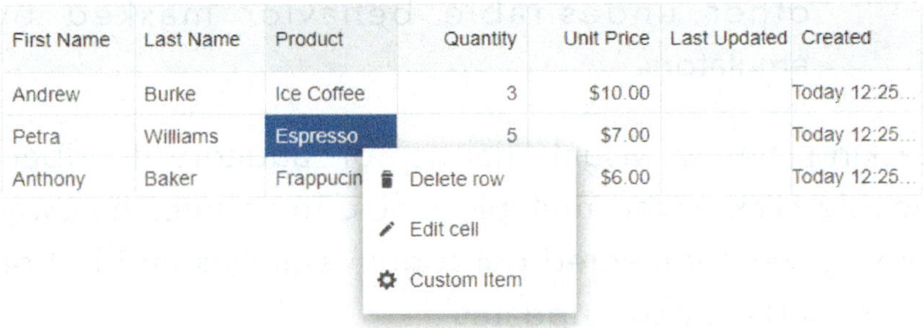

Figure 39: Context menu - htmlelements.com

Steve Jobs famously hated multiple button mice (macrumors.com article from March 11, 2014). However, capacitive touch sensors in today's minimal "Magic Mouse" saved the day and allowed Mac enthusiasts to activate context menus with a right click gesture. In contrast, people commonly used multi-button mice on Windows sometimes pushing limits of usability.

Other details contribute to elegance such as clear, concise, and informative explanations in dialog boxes. Seek out your documentation team if this skill eludes you.

Design teams typically set layout rules unless you are building on a foundation where these already exist. Inconsistent spacing, layout and typography ruin the impression of an otherwise stellar interface.

Efficiency and robustness also contribute make or break aspects of elegance. Do people experience delays after clicks or button presses? Immediate feedback of clicks or taps counteracts latency introduced by mechanisms.

Test with real hardware to expose delays and other undesirable behavior masked by emulators.

Lacking timely visual, haptic, or auditory feedback, people click or tap multiple times. To counter hardware latency, we engineered our toggle switches and buttons for immediate visual updates.

Feedback takes the form of a pop up (toast) or with a visual change, such as immediate toggling. (Figure 40) Only if the action failed did the toggle switch return to the previous on or off state with appropriate failure feedback from a toast popup.

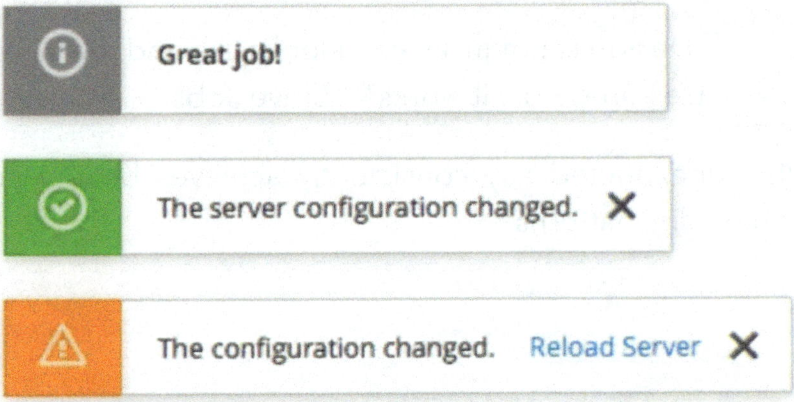

Figure 40: Sample Toast Messages — sheribyrnehaber.medium.com

Long delayed actions require some type of busy indicator. Better yet, provide progress indicators so people have a good idea how long the action will take. (Figure 41)

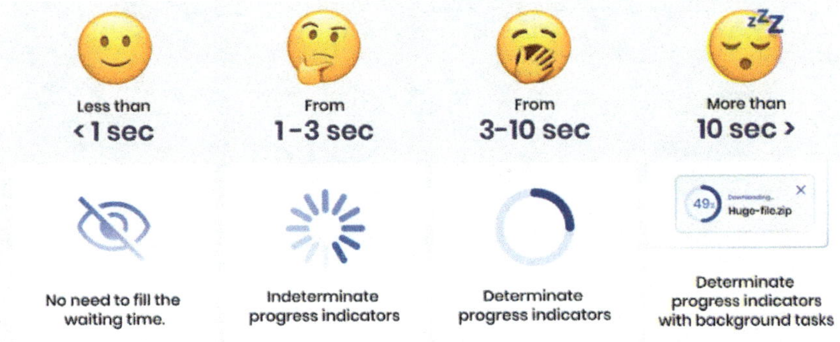

Figure 41: Busy vs Progress Indicator — uxdesign.cc

Your audience senses responsiveness and fidelity of action in determining quality. Bugs, crashes and freezes truly frustrate people, regardless of reason. In Chapter 8: "Avoiding Disasters," we discuss reasons for such misbehavior.

"Design is not just what it looks like and feels like. Design is how it works" – Steve Jobs.

What unexpected ally consistently achieves beauty and high quality patterns?

CHAPTER SIX

Building Solid Foundations on Frameworks

"Everyone loves to create ideas but fewer build upon solid foundations of thought"
-- C. A. A. Savastano, Writer about Democracy

In the 1980s, modular object-oriented architecture remained theoretical due to tiny memories and rudimentary software tools available at that time. Developing software for embedded machine controls involved arcane tools, day long software builds and ponderous debugging. We developers routinely slogged through computer hexadecimal memory "dumps" scanning for clues left by our misbehaving code. (See Figure 42). Object-oriented software and concepts of frameworks were more of an academic exercise than an industry standard.

```
$hexdump samplefile.txt
0000000 6548 6c6c 2c6f 5720 726f 646c 0a21 6854
0000010 7369 6920 2073 2061 6574 7473 6620 6c69
0000020 2e65 310a 3332 3534 3736 3938 0a30 6854
0000030 2065 7571 6369 206b 7262 776f 206e 6f66
0000040 2078 756a 706d 2073 766f 7265 7420 6568
0000050 6c20 7a61 2079 6f64 2e67 4c0a 726f 6d65
0000060 6920 7370 6d75 6420 6c6f 726f 7320 7469
0000070 6120 656d 2c74 6320 6e6f 6573 7463 7465
0000080 7275 6120 6964 6970 6373 6e69 2067 6c65
0000090 7469 0a2e 6553 2064 6f64 6520 7569 6d73
00000a0 646f 7420 6d65 6f70 2072 6e69 6963 6964
00000b0 7564 746e 7520 2074 616c 6f62 6572 6520
00000c0 2074 6f64 6f6c 6572 6d20 6761 616e 6120
00000d0 696c 7571 2e61 550a 2074 6e65 6d69 6120
00000e0 2064 696d 696e 206d 6576 696e 6d61 202c
00000f0 7571 7369 6e20 736f 7274 6475 6520 6578
```
Figure 42: Memory file hexadecimal output - tutorialspoint.com

Moore's Law advocated doubling computer speed and memory every two years. This drove relentless progress by exponentially increasing transistor count and thus, computing capacity. Less than a decade later, developers embraced reusable "object-oriented" software necessary for frameworks, and new software tools automatically converted memory for easy reading. Breathtaking vistas of never before imagined personal computers and software opened up.

Improving software and hardware technology yielded products like the original Macintosh. My enthusiasm for the "Mac" and its graphical interface resonated with veteran engineers and helped land my first post-college job. The company coincidentally emerged from its primordial punched card era.

Proving myself over the first years meant contributing to the existing scratch-built operating system (see Figure 43). While admittedly crude, no industry standard OS existed for real-time machine control, let alone one supporting a graphical user interface. With our scratch built OS, implementing a drop-down menu meant custom building code. Our team took wobbly baby steps during those early years toward my dreams of attaining the Mac's iconic beauty.

Figure 43: Our machine control before GUI - exapro.com

At the same time, Microsoft responded to the Mac by quickly following up with Windows. We saw original Windows as a toy compared to our stringent industrial requirements. As hardware technology quickly matured, Microsoft took its moonshot toward supercomputing with Windows NT.

Maturing microcomputer technology emboldened us to replace our outmoded operating system with an industry standard. The chief engineer favored IBM's OS/2 over Windows NT, but the rest of us engineers unanimously overruled him in favor of Windows NT. As we correctly predicted, OS/2 never gained traction against the juggernaut of a better supported Windows developer community. It may be rightly argued that times have changed since then!

Even with Windows as our OS, we still needed to write our own touchscreen-friendly widgets. Machine shop applicability again required crafting specialized widgets (see Figure 44). For example, the tool manager application showed tools both in graphical and textual form for the first time in a machine tool user interface.

To a skilled machinist, there was an almost visceral feel to the data we presented. This rich graphical experience extended throughout the user interface (more examples later). Visual frameworks, like ours, revolutionized software development. Even so, Jonny Ive, Apple's minimalist design champion, would likely cringe at our 1990s skeuomorphic aesthetic!

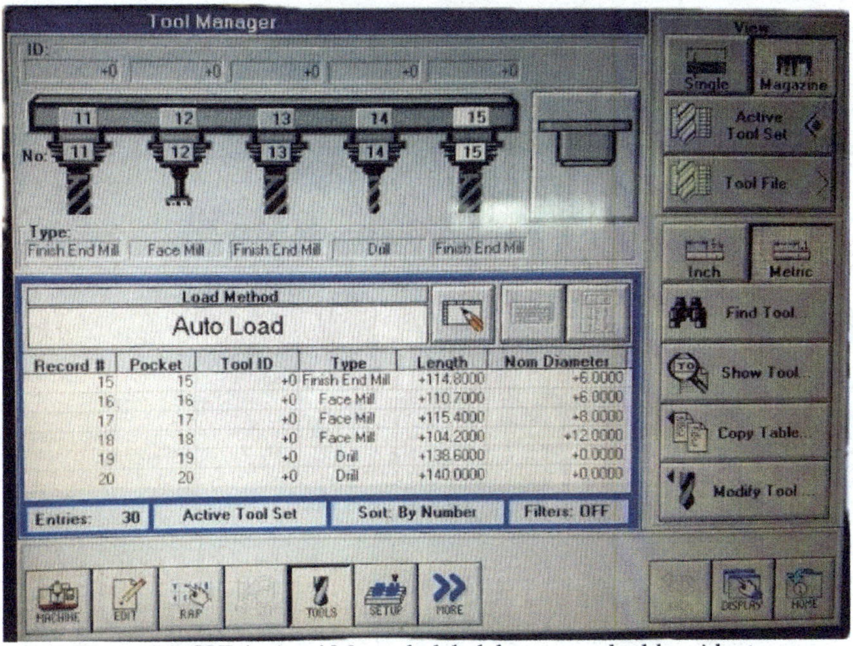

Figure 44: GUI Arrives! My early label, button and table widgets - maszyneria.com

In my career's second phase, my colleagues designed and implemented a service infrastructure that blended efficiency, portability and usability. Web technology to the rescue?! Initially, most web frameworks served up static data at web page load time (REST interfaces). Instead, we purpose-built our own Web Sockets framework that triggered events on demand, avoiding wastefully polling for data.

This design, patented by my colleague James, featured real-time updates suitable for battery powered tablets on erratic warehouse Wi-Fi networks. By design, our framework achieved efficiency, durability, massive real-time throughput and low power consumption. Writing a large multi-faceted application on top of an existing widget library required minimal application code to achieve dazzling on-screen responsiveness

(See Figure 45). I reveal our accompanying resource pitfalls in Chapter 8: "Avoiding Disasters."

Figure 45: Example modern widget library — Telerik.com

After our acquisition by a Fortune 500 company, we redesigned the UI using open source and extended the corporate widget library with our own features. Due to the framework's mission criticality, we used unit tests as the first line of defense against breaking changes, but such tests prove only as good as their coverage. We also enforced painstakingly detailed reviews before checking in code.

Frameworks require strict code management and exhaustive unit tests.

That said, many platforms feature all you need to build a high quality product with off-the-shelf widgets regardless of event driven or polled. Indeed, iOS and Android frameworks thoroughly support most applications. However, if your visual framework proves unfit for your requirements, investing in at least some of your own widgets and services helps deliver excellence to your marketplace.

Products needing to last for decades need resilience beyond "good enough".

Framework layers standardize various aspects of software implementation. The list below includes a few of many potential functionalities that a framework can provide and enhance:

1. Database abstraction: to separate business logic and package data in display form.
2. Multiple computer data collection, aggregation and distribution: to offload UI processing from machine controllers and combine data sourced from distributed multiple controllers.
3. Localization: support dynamic language switching.
4. Layout: eliminate sloppy spacing variations for

pixel-perfect positioning. Support standardized dialogs for both save and confirmation.

5. Unit Conversion: automatically convert to an individual's preferred measurement system.

6. Currency Conversion: convert local currency to make international customers feel welcome.

7. Brand Styling: establish familiarity and loyalty to your company.

8. Theming : adapt to time-of-day lighting conditions or directly choose light, dark and branded elements.

Recently, Apple's operating systems introduced "Liquid Glass" across their product lines. The aesthetic proved quite controversial. Some like the shiny glass and others strongly oppose it. Worse, transparent elements degrade legibility, especially for those experiencing visual contrast issues. Because Apple's iOS uses a framework, the software team implemented an easy-to-add opacity adjustment and addressed the accessibility issues.

At first, Apple hid Liquid Glass transparency settings under Accessibility settings, but later addressed legibility concerns by moving the transparency setting to the forefront. Now everyone wins by deciding how much or how little "Liquid Glass" transparency works for them (Figure 46). A company must merely dedicate itself to good architectural principles and avoid forcing a controversial experience on everyone. If only people settled typical disputes this easily!

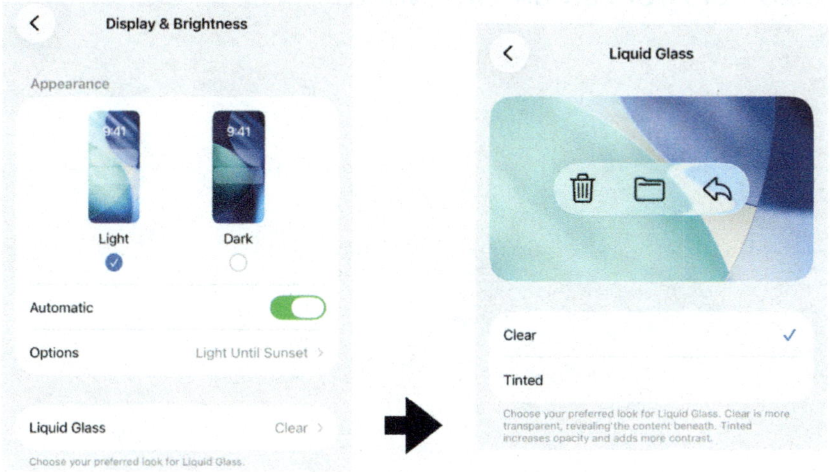

Figure 46: Liquid Glass democratized (easy display setting) — Apple, Inc.

Apple's controversy illustrates one facet of how frameworks contribute to your infrastructure. The framework changes did not "break" the UI in the conventional sense, but voices of reason wrought easily implemented compromise.

Update your high quality widget library instead of updating applications piecemeal.

AI frameworks now offer another fundamental shift. These frameworks already deliver AI editing tools for text box widgets built on the framework. Deeper integration of AI with supplied contexts promise greater assistance throughout consumer and industrial devices.

So far we've explored a number of tools for building a better experience, but how do you get the most demanding customers to love your product and draw customers from around the world?

CHAPTER SEVEN

Personalizing Appeal in General Purpose Software

"The objects you own should serve you - not the other way around.."
-- Marie Kondo, Organizing Consultant

The UI holds complete responsibility for software that adapts to different roles, streamlines common workflows, and works internationally. Such customization features make the difference between world class software and everything else. I helped evolve the design and implementation of such increasingly capable features across decades of development.

My first major instance of supporting customization arose from engineering a touch screen machine control a decade before the iPhone's multi-touch interface. Our UI thrilled our customer base. However, a single feature stood out. People wanted a "Home Screen" to conveniently access their favorite features. We implemented a mode for users to assign Home Screen buttons by simply:

1. *Activating Home Menu configuration.*

2. *Navigating to the appropriate screen.*

3. *Pressing the button to assign.*

Behold! The UI returned to the Home Screen and the icon and label magically transformed the blank or pre-existing button function to the configured one. People talked about their fondness for this feature even years later.

More recently, our home screens and UI configurability by role drastically improved. We presented each role with a customized experience and each person could further customize their own dashboards. For almost a decade prior, designs for dashboards were frequently on my mind as an ideal. Realizing that dream provided another career highlight. Better yet, we were brainstorming enhanced dashboards directly with great minds on the customer side. (See Chapter 1).

Even before that, we included home screen functions for starting up and shutting down production each day (operations) or specialized maintenance functions. Different industries require their own special processes and procedures.

Sometimes your audience consists of multiple groups of individuals such as:

- Project Engineers setting up a machine
- Specialized maintenance teams
- Minimally trained operators

Roles and customization present appropriate features to the currently authenticated user.

Settings, Home Screens and Dashboards delight power users.

Customization appeals to the most demanding and exacting of people's wants and needs. People expect more personal experiences especially from luxury brands. For industrial UIs, such features differentiate industry leaders from their competitors.

Settings achieve first level customization. If your audience disagrees about how to best express and prioritize vital features, provide settings and persistently store them (best case, with cloud backup). People have high expectations to preserve settings because setting up a new phone allows them to pick up exactly where they left off with the previous one.

Settings also prove ideal for features that detract from the experience of many users, but others must have. Primary examples include accessibility features like large text, high contrast, emphasizing audio dialog over ambient sound, addressing of motion sickness, etc.

An HMI Builder delivers top tier product customization to your most demanding customers. Ideally, produce a highly customizable application that users build on or modify to their liking. Most customers only build their own UX when believing themselves practitioners of UI design or desperately fulfilling highly specialized needs. The worst case scenario involves customers building bad

UI and blaming you for their own bad result. Potential loss of partnering for upgrades further complicates reliance on third party tools.

In most cases however, home screens offer the appropriate solution. Many home screens host customizable dashboards. Below is a Home Screen example from the Apple Home app. (See Figure 47)

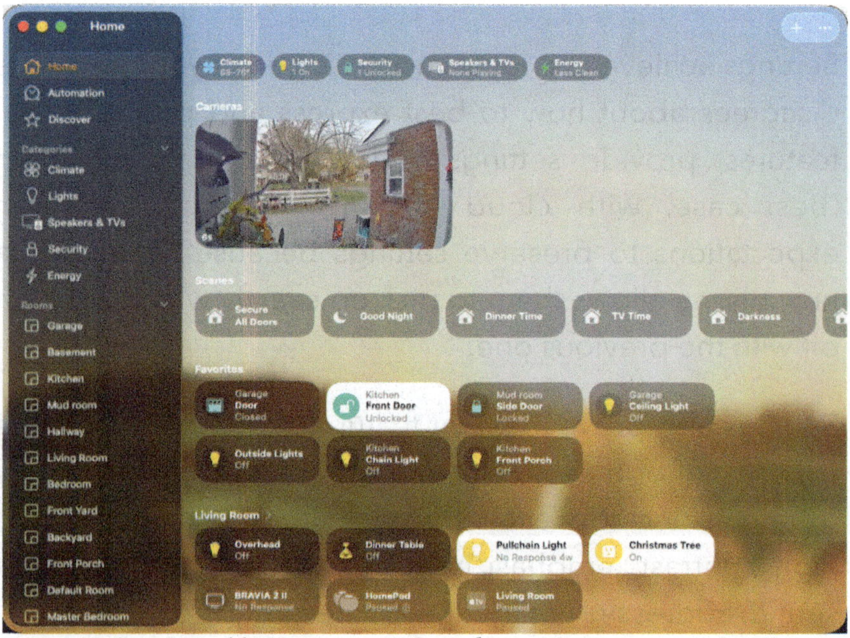

Figure 47: Customizable Home screen from the Apple Home app - Apple, Inc. - author

Cars and trucks offer increasingly extensive configuration as they become more sophisticated "smartphones on wheels". People may also prefer neutralizing certain features added on their behalf such as a key fob unlocking only one door (as opposed to all) or overly sensitive lane departure alerts. Do you want your car popping up a surround view at each intersection while trying to navigate a GPS route? I experienced such an annoyance following a software update for one of our cars.

On a more encouraging note, two examples of "Goodbye Screens" stand out. Power off automatically pops up handy access to features for charging the car, pre-climatizing and opening the hatchback before you walk away. Tremendously useful! (See Figures 48)

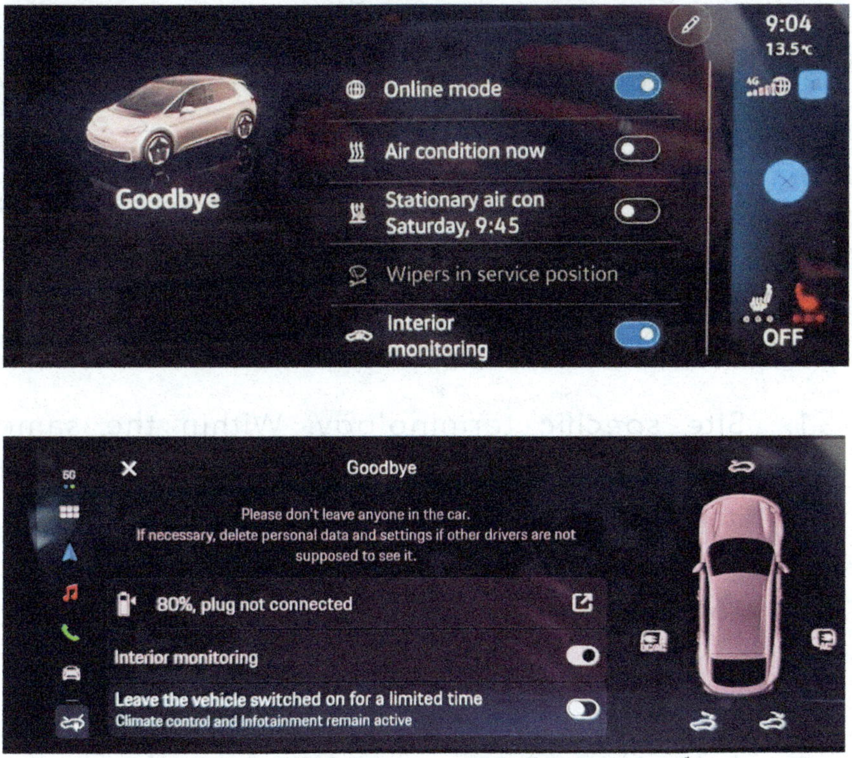

Figures 48: Goodbye Screens - vwidtalk.com and author.

Providing highly configurable dashboards levels up the user experience. Power users delight in such features especially when built with high quality widgets. Configuration of the dashboard may include units, levels of detail or choice of data to include or exclude.

Provide widget defaults to allow easy and immediate use, but also configuration for power users.

Running a machine typically requires a startup sequence and running all that from a single screen improves the experience over requiring multiple screen navigation. Before customizable dashboards we heard many complaints of too many clicks.

Companies also look to globalization to extract additional profit from their products. Supporting localization and/or internationalization accordingly customize your creation. Sample opportunities here include:

1. Site specific terminology. Within the same language, there may be site specific names of items in the environment, such as conveyor lanes, guided vehicles, robots, cutting machines or conveyor lines.

2. Limited space for translated text.

3. Communicating context to translators for precise translation.

4. On the fly language switching (for the sake of tech support).

Effectively supporting internationalization requires engineering your product with:

1. Adequate and flexible spacing for text. For example, Portuguese may require 150% of space needed for English

2. Supporting formats like XLIFF. The XML Localization Interchange File Format standardizes sharing your translatable content with translators.

3. Tooling and services to host and switch between multiple locales. This may require language packs and other setup mechanisms. Database data further complicates support for multiple languages. Choose to express translated text or embed multiple languages in the same column. We created a special language string type allowing the client to dynamically change between a few languages. This type provided all translations in one container.

4. Learning translation support software and using provided file exchange tools. Excellent software tools exist for collaborating with translators (see below). I also recommend taking time to provide translators with "in context translation" resources such as screen shots.

Leveraging tools like Alchemy's Catalyst and in context translation set you up for success. (Figure 49) Don't try to build your own tools. Existing translation tools offer decades of enhancements and details you probably would miss. I better appreciated the rich set of testing and translation tools in Catalyst as I learned to leverage its capabilities.

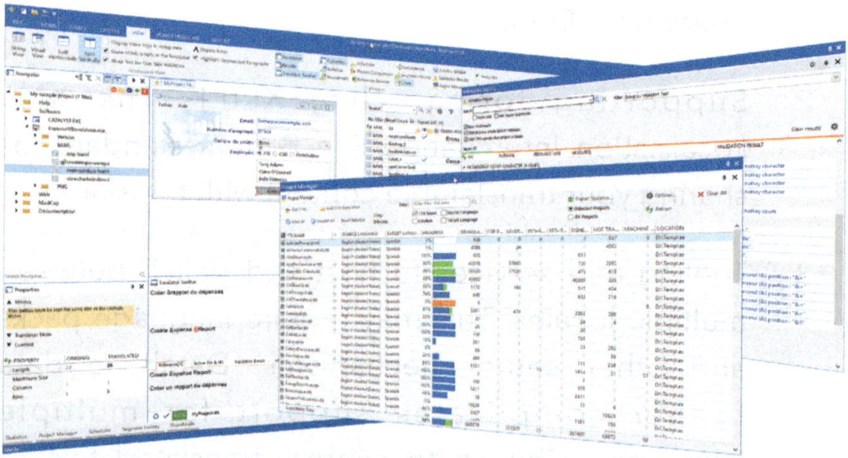

Figure 49: Alchemy Catalyst screens - alchemysoftware.com

Be aware that many translation vendors may prefer hosting online translations rather than supporting embedded and offline products like control systems. Standalone systems can't receive updates from remote servers. Provide the right infrastructure for internationalization up front and include testing! Don't underestimate the complexity and importance as you will find retrofitting localization very time consuming and costly.

Choose to buy from translation tool vendors appropriate to your requirements instead of making your own tools.

Catalyst's in context translation tool allows translators to visualize how each phrase appears on screen, but translation resources require setup and maintenance. Catalyst eliminates tricky testing that normally requires painstakingly navigating the UI and simulating multitude conditions. Translators simply call up the text and see it as it would appear in the UI. Phrases drive navigation rather than demanding full function testing.

Shoddy AI translations ruin otherwise solid products and may embarrass you by being unintentionally funny and/or humiliating. Though translators leverage AI and glossaries for the first pass, selecting exactly the right translation remains a fine art. Only those who are multi-lingual fully understand translation pitfalls.

Target cultures may even laugh at your brand name. For example, the original phonetic translation of "Coca-Cola" into Mandarin characters roughly sounded like nonsense akin to "Bite the Wax Tadpole". A minor adjustment yielded the more appealing pronunciation of "tasty and fun" (see snopes.com).

While teams incur a relatively small risk of outright failure due to translation issues, poor translations increase need for support and muddy an otherwise stellar UI. Therefore, involving people from your target culture promotes favorable outcomes and may even improve the core product. Improvement often results from exposure to different perspectives.

Adding a fluent multi-lingual speaker from your target culture to your team drastically improves first time localization.

Powerful customization and precise localization reveal well designed products. What missteps may critically damage a well designed and possibly already successful product?

CHAPTER EIGHT

Avoiding Disasters

"The job of testing is not to find bugs but to find features that don't meet the user's needs"
-- Michael Bolton, Leader in Context-Driven Testing Methodology

Many books explore Quality Assurance (QA) in generous detail and this book avoids such a deep dive. However, I document here my hard won lessons about quality threats in hopes of steering you clear of a potential shipwreck.

As stated in previous chapters, business leaders often trivialize the UI and system engineers frequently provide the UI whatever is left over of scarce system resources. Poorly provisioned virtual machines (VMs) lead to unexpected failures. In this situation, your UI exhibits behaviors such as “freezing” and memory related crashes (Figure 50).

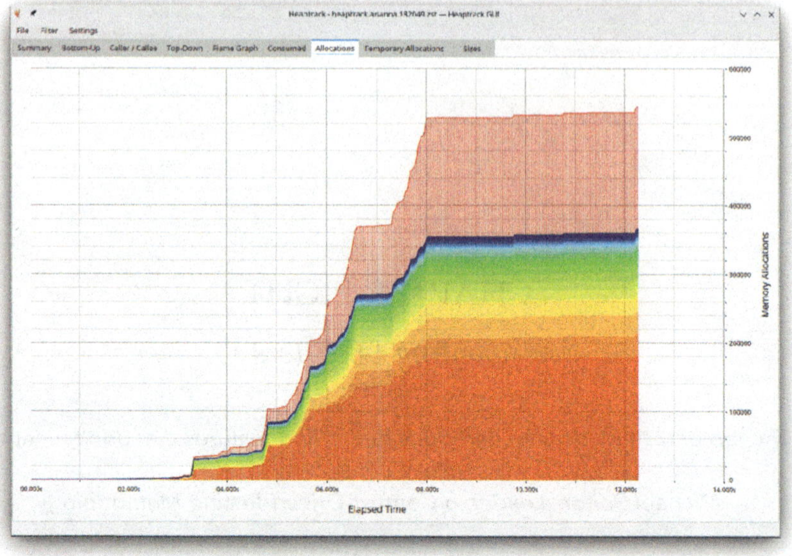

Figure 50 Memory usage over time — apps.kde.org

Having our UI installed on inadequate customer VMs resulted in many support headaches. I was amazed by how many ways our project engineers and customers accidentally deprived the UI of memory, storage or processor (CPU) resources it needed to work well. Customers ran at lower capacity during commissioning and acceptance testing, thus keeping stress related failures concealed. During later peak operation, capacity issues showed up and with greater urgency to resolve.

As an antidote, I exhaustively documented server and client requirements for browser, memory, CPU, and storage. In our case, we sought dedicated processor cores on a multiple core CPU chip. Having our own core prevented demanding real-time processes from consuming most of the available time due to the UI's lower assigned priority.

Following a pattern found in both mobile devices and computers, I added icons to make low resource conditions evident to the top right application main bar. Evidence of errant conditions would appear in any screenshot that our team received. We named the software feature "Resource Manager" and dedicated it to colleagues in customer service who commonly dealt with these recurring irritations.

Frustratingly, we often could not test at scale and it took many hours of trace analysis to pinpoint root causes as typically caused by various limited resources. This meant that it took years to convince colleagues to look first for resource issues as the most likely cause of UI problem reports. A startup or smaller business might fold before achieving suitable reliability.

Your customer's environment is not kind to your software. Perform brutal stress testing.

Attention to the following details should help to avoid similar discomfort to mine:

- Sufficiently understand and extensively document the resource requirements of your UI.

Windows Server Version	Processor	Memory	Storage	Recommended
2016	1.4 GHz x64 processor	1 GB Server Core 2 GB for desktop	32 GB (minimum to install)	2 GHz x64 8 GB RAM 100 GB Storage
2019	1.4 GHz x64 processor	1 GB Server Core 2 GB for desktop	32 GB (minimum to install)	2 GHz x64 8 GB RAM 100 GB Storage
2022	1.4 GHz x64 processor	1 GB Server Core 2 GB for desktop	32 GB (minimum to install)	2 GHz x64 8 GB RAM 120 GB Storage
2025	1.4 GHz x64 processor	2 GB server Core 2 GB for desktop	32 GB (minimum to install)	2 GHz x64 16 GB RAM 160 GB Storage

Figure 51: Sample Resource Requirements — author

- Make glaringly evident situations you can't control such as resource issues (low memory, low storage and high CPU utilization). The "Resource Manager" documented above provided our solution. Noticing such issues in a screenshot was guaranteed by badged icons in the system title bar of the application, but you may choose even greater emphasis.

Figure 52: Low Resource Warnings — news.trendmicro.com

- Test for problem cases such as low bandwidth or frequent network disconnects. In other words, purposely try to break your software by running it in environments simulating bad networks, low memory and low CPU bandwidth. Achieve this by decreasing cores available, filling up available memory, entering and exiting wi-fi range or using a device that can simulate network disconnects.

Figure 53: Low Bandwidth — liquidweb.com

- Create metrics and automated tests to proactively check for performance degradation and resource leaks. While developers must manage memory for languages such as C++ making errors more likely, garbage-collected languages like C# and Java may also leak. Be aware, too, that other resources besides memory may leak! Usually these manifest as memory consumption but could also cause performance degradation. Free storage space also presents a potential pain point especially when logging excessively.

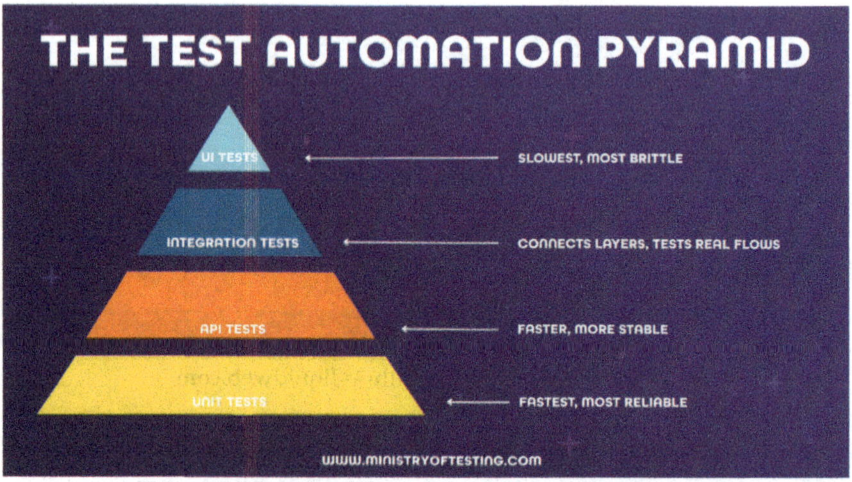

Figure 54: Levels of Testing — ministryoftesting.com

- Test using realistic hardware and environments. Frequently developers test with simulated or emulated environments, but you must eventually create realistic test environments or risk embarrassing or disastrous failure at company sites. Do your best to personally run equipment as your audience would. From my own experience, you almost always learn lessons from spending time running your system and experience exhilaration from what you made, too!

Figure 55: Example Test Environment — jpl.nasa.gov

Pre-release testing requires realistic environments, including simultaneous users, high capacity, CPU load, etc.

Sometimes you can't test against reality, but closely approximate real world environments with stress testing as a substitute. This is easier said than done, but if you don't do it proactively, you end up doing it reactively to address escalated issues at customer sites. From personal experience, avoid customer site instability at all costs!

Your best friends in QA treat your software mercilessly.

You are well on your way to building a world class product, but how do you protect your hard work?

Part 2: Safeguarding Your Product and Career

CHAPTER NINE

Protecting Intellectual Property

"Patents are like fertilizer. Applied wisely and sparingly, they can increase growth. But if you apply too many chemicals, or make patents too strong, then you can leach the land, making growth more difficult"
-- Alex Tabarrok, Canadian Economist

I admit, patents embolden some and enrage others. Regardless, please be aware of patents and their use at least as an opportunity. If you rely on open source, you may agree with Richard Stallman who said "Software patents are dangerous to software developers because they impose monopolies on software ideas."

However, James Dyson claimed "If you didn't have patents, no one would bother to spend money on research and development. But with patents, if someone has a good idea and a competitor can't copy it, then that competitor will have to think of their own way of doing it. So then, instead of just one innovator, you have two or three people trying to do something in a new way."

Patents must be novel, but novelty may merely extend an existing patent or involve application to a different discipline or industry.

For example in 2004, I patented the idea of associating bar codes with configuration data. People could setup a newly purchased machine tool accessory without entering any data. System software associated barcodes with downloading and updating configuration data.

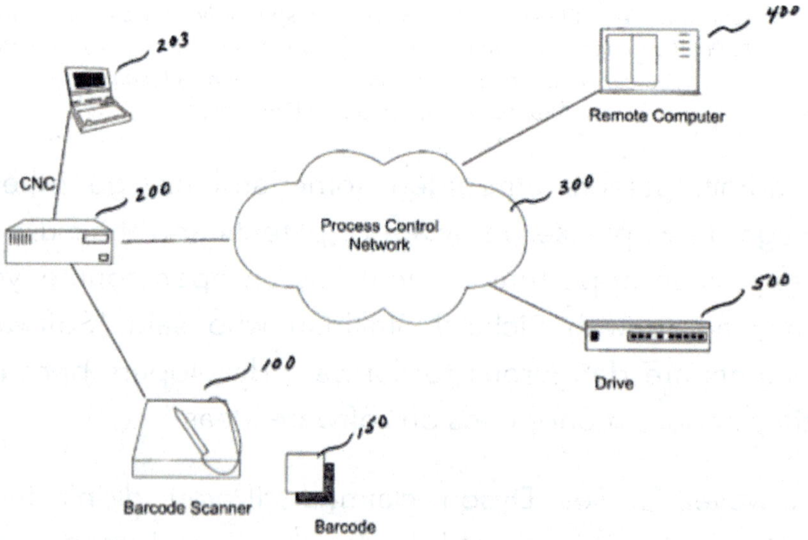

Figure 56: Patent diagram for US 2004/0222300 A1

Associating barcodes and data offers no special novelty, but the United States Patent & Trademark Office (USPTO) considered barcode-triggered machine setup novel enough to grant a patent.

You can make your own choice about protecting your intellectual property, but if you don't patent, you may spend a lot of time and money on innovating only to have someone else quickly copy and market your idea.

Here are some realities about patents in general and software patents specifically.

1. You can't patent software per se, but you can patent processes.
2. A novel idea is patent worthy only if it is not common sense, and no prior record exists of the idea being patented.
3. A patentable idea must only be novel in application to a particular industry or specialty and not necessarily novel overall.
4. Claims describing the novelty of an invention provide the key and essence of a successful, enforceable patent.
5. You need to involve a patent lawyer. If you work for a large corporation, your employer may supply services of a patent lawyer. Otherwise find a patent lawyer with software experience. These services cost upwards of $15,000.

We've examined ways to protect a quality product, but what about necessary qualities of your employer?

CHAPTER TEN

Identifying Desirable Employers

"The first responsibility of a leader is to define reality. The last is to say Thank You. In between, the leader is a servant"
– Max de Pree, management philosophy and human-centered design leader

Great products and partnerships arise in supportive organizations. Sometimes you "make your own luck" by selling your company on a quality user interface, but such efforts end up difficult to impossible in toxic work environments.

From 1984 to 2012, I worked with exceptional teams mentioned in the dedication. My first manager Tom led with heart, but I went through some rough patches afterwards until joining Jim's team. Jim and I started working together as brainstorming collaborators and both delighted in the design and implementation phases of projects. Jim catalyzed both my self esteem and career, eventually leading to a management role and providing mentorship throughout. Working with someone like Jim seemed to me like a peer collaboration with Jonny Ive. Such experience makes even a mega corporate job worthwhile. If such an opportunity comes your way, take it!

Putting my opportunity into perspective, Jonny Ive was Apple's industrial design chief who advanced the state of the art and overall quality of Apple Products in lockstep agreement with Steve Jobs. In our case, Jim made our signature UI a key selling point when a mega-corporation acquired our team. Without warning, new managers decided against the original plan to integrate our UI. We successfully took on new roles for almost a decade, but the 30+ year collaboration of our talented team was eventually phased out. A few years later some of us found new opportunities in a different industry as described below.

My second great opportunity involved a phenomenal material handling company. The first four years there reinvigorated my career. In late 2016, I spoke up at a farewell meeting for our founders to recognize and honor their contributions to corporate culture. I remembered the company president for vocally appreciating the painstaking craftsmanship built into the company's products and how these words motivated new hires to sustain already existing momentum. Leadership also showcased engineering contributions in employee and board room meetings.

Without grasping the full consequences, our founders sold us to a mega corporation with a different culture. We reluctantly witnessed how new management systematically replaced the founder's values and abandoned the company's proudly built up manufacturing sites. The abandoned sites remind me of the disintegrating Lumon factory shown in Apple TV's series Severance.

Building a great UI requires leaders that grasp the UI's mission-critical contribution to success.

I once heard a corporate visitor dismissively say, "this is the problem with founder run companies" referring to our independent spirit. Not surprisingly, our customers noticed the changes and missed the personal attention of our founders resulting in losing business to competitors.

The genius of a "solutions" based company means customers concentrate on their business model while leaving automation and warehouse construction to a partner with painstakingly cultivated and complimentary subject matter expertise. Our purchaser's product expertise meant systematically sacrificing the subject matter experts to corporate dogma. Their leadership seriously believed that franchise formulas and process yield better profits than expertise.

We soldiered through franchise mandated "Codes of Conduct" and "8 Behaviors". Corporate leaders believed people better motivated by punishment than reward, thus squeezing out appreciation for employee innovation. It did not take a prophet to read the signs. Our immediate supervision provided expert insulation from some of these corporate trappings, but such organizations made their efforts an uphill battle. (Figure 57)

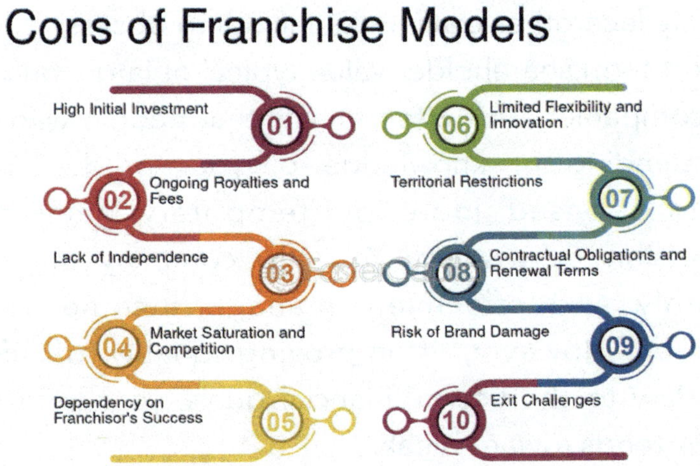

Figure 57: Risks of franchise models — fastercapital.com

Mega corporations frequently sell process internally because evangelizing process justifies many jobs at corporate headquarters. You have to argue against or submit to Six Sigma, Failure Mode Effects Analysis (FMEA), grueling security procedures and exhaustive performance metrics. In contrast, our founders favored minimal process typical of our more nimble competitors. Startups, like the company I originally joined, disrupt industries largely by avoiding process-related obstacles. You're headed for ruin when corporate leaders see process as the product.

As a postscript: in April of 2026, the mega corporation sold what remains of our company to an investor group for considerably less than the purchase price. The same investor group also owns a former competitor where many former colleagues flocked. This acquisition reunites numerous colleagues separated by intentional departures or layoffs. Exit challenges (Figure 57) remain, such as unwinding IT infrastructure and other dependencies.

You may face other challenges including obsessive focus on short term shareholder value typical of large, publicly-held companies. This often means that people with little understanding or knowledge of your product make decisions based more on temporary gains than compassion or wisdom. Yes, the company must run smoothly and effectively manage income versus expenses. However, if improving quality of life is secondary to short term financial goals, your long term outlook tends toward bleak.

Producing a user interface for a significantly complex product such as a phone, car, industrial machine or material handling system requires skill and determination. If your company considers the UI trivial, you may expend major effort to dispel such notions. To counter notions of UI triviality, I created a presentation detailing the size and complexity of our UI. The slides spelled out all the disciplines involved, the different user roles and the sheer immensity of the data model. Hopefully you can use this book and provide similar reasoning to dispel misguided notions in your workplace.

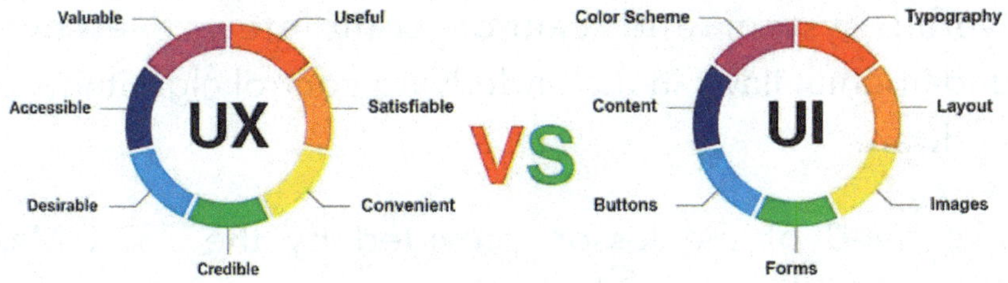

Figure 58: Diagram of UX and UI Design Elements — leosiraj96.medium.com

Customers often neglect usability in purchasing decisions, unless they include representatives from their operations or maintenance teams. This is why the benefits of good designs need prominence in marketing materials and wow factor demonstration during sales presentations. Provide demo software and encourage customers to perform common tasks. Given an elegant product, demos showcase marketplace advantages and allow the UI to sell the overall product to customers.

If you work in companies with low people priorities, you must slog uphill to evangelize importance of the user interface. Your customers already know the value of the UI and hopefully help you convince their business leaders and yours. However, frequently dismissed voices of reason may lead to better applying your talents elsewhere.

Also, be especially wary of situations where the UI must strive to overcome an otherwise bad design or poor configuration. Some amusingly refer to "putting lipstick on a pig", since an effective UI must participate in a solid overall system and cannot completely overcome fundamental flaws in the underlying control algorithms or hardware.

Take heed of the lesson provided by the 737X Max. (Figure 59). Corporate leadership imagined that software could overcome the airplane's more challenging flight dynamics, but such notions proved tragically false (corpgov.law.harvard.edu). If your UI is part of a system with fundamental flaws, you may temporarily hide the shortcomings, but poor design shows itself at the worst possible time. Beware of those blaming the UI or other software for failing to magically fix underlying flaws. This foreshadows peril for both you and your customer.

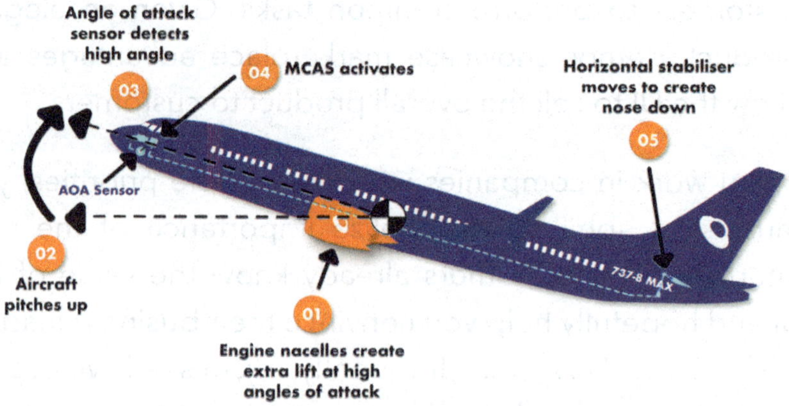

Figure 59: Boeing 737 Max design issues - pilotswhoaskwhy.com

Hard lessons from my management experience convinced me that companies substantially reflect the personalities and quirks of their leaders. Some companies now set impossible goals. Business leaders can't overcome physics, inadequate data or a fundamentally bad design. SpaceX finally backtracked on the impossible goal of sending humans to Mars before becoming proficient with Earth's moon. However, too many businesses over-promise and under-deliver, whereas the most valued businesses under-promise instead.

Other major risks arise out of management decisions to save money. AI vision systems lack enough sophistication to differentiate between a living or damaging obstacle and a plastic bag. When overly relying on bad decisions such as optical-only Full Self Driving (FSD) has already resulted in devastating consequences. Competing systems use LiDAR for redundancy and these enhance human senses rather than try to be "good enough". Faustian bargains frequently arise when greed beats out quality and concern for people.

Recognize a "Faustian Bargain" when you make one.

Consumer products companies like Apple or Samsung make selling UI advantages easier. However, less directly consumer-focused companies (like industrial products) obscure benefits of a well designed and implemented UI. I created and maintained UIs for machine tool and material handling industries. In such non-consumer focused industries your results alone won't speak for themselves.

For comic relief, here are the so called "Six Phases of a Project"

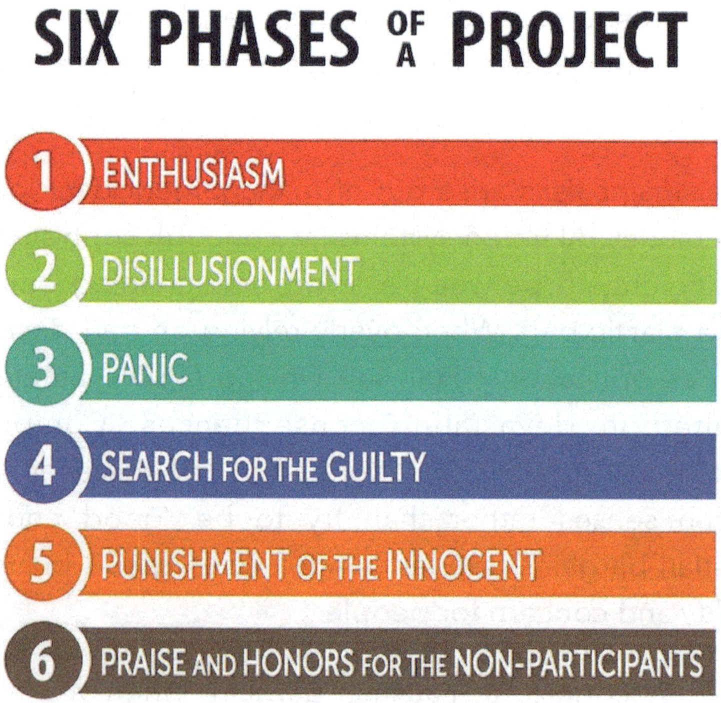

Figure 60: Six phases poster — cafepress.com

Also, remember the wisdom of Fred Brooks' mythical man month. You can't deliver software faster simply by throwing more people at it, especially towards the end. I've seen this tried many times and always fail spectacularly.

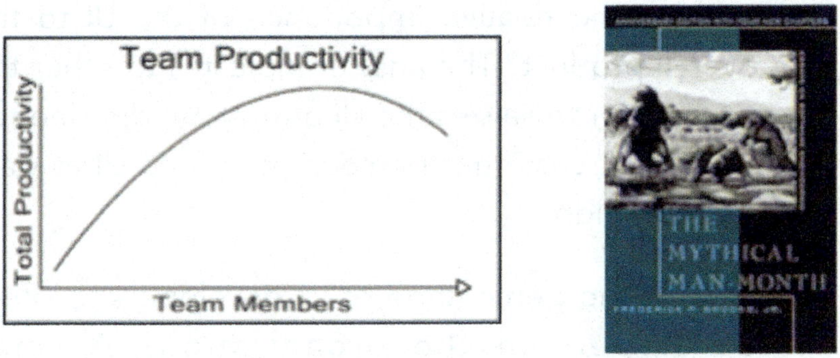

Figure 61: The Mythical Man-Month effect and book cover — thegraphblog.wordpress.com, bookshop.org

Carefully determine the situation you are hiring into. Are company leaders wise, solid and supportive of their employees or only in it for maximizing profit at the expense of everything else?

Working for leaders like Tom, Jim, and Rich (mentioned in the dedication), act as a needed buffer between dubious corporate bureaucracy and their teams. Caring leaders identify and cultivate talent, making their team's contributions feel vital.

Depending on the job posting relative to your personal interests and aptitudes, here are some questions to ask:

1. What do you look for in a member of the UI team? The answer should include more than just coding skills. Candidates should effectively communicate both verbally and graphically. For the foreseeable future, imagination, curiosity and technical aptitude hold high importance.

2. What is the relative importance of the UI to the overall product? The answer indicates whether the interviewer trivializes the UI or not, but dig deeper. Ask about customer perceptions of the UI to gain further insight.

3. What is the percentage of UI developers to other developers in the organization? A small percentage or small overall number warrants explanation.

4. What do you look for in lead team members? Unless interviewing for a lead position, this tells you if your skills line you up to advance and qualifies your interviewer also. Such technical leads set examples with technical abilities but also need to mentor successors for future team viability.

5. What does the corporate culture value most? An answer like "growth" flashes a danger sign unless it means personal growth rather than strictly earnings growth. Look for something like improving lives or innovation. For answers like "innovation", ask clarifying questions.

Don Norman calls on us to select companies building responsible products in his book *Design for a Better World*. Such products combine sustainability, robustness, resilience and materials that self regenerate. Approaching design with a steady eye on environmental consequences challenges us for decades to come.

Figure 62: Don Norman's *Design for a Better World* — mitpress.mit.edu

Finding responsible employers may seem impossible when companies seem more interested in maximizing profits by replacing employees with AI and robots. CEOs proposed mass job replacement throughout my career, too. I hope you can align your personal vision with a potential employer's corporate culture, but certainly understand making compromises.

Next, how can we understand the product life cycle to avoid business decisions ruining an otherwise successful product?

CHAPTER ELEVEN

Navigating Change and Its Obstacles

"The people who are crazy enough to think they can change the world are the ones who do"
-- Steve Jobs, CEO of Apple, NEXT and Pixar

Carefully educating potential buyers about how your new product benefits them presents a challenge. Not every product has the gravitas of a highly anticipated Apple product. Steve Jobs expertly drove home the point of iPhone when he repeated: "a revolutionary mobile phone, a widescreen iPod, and an Internet communicator." He then asked: "Are you getting it?"

Figure 63: Steve Jobs Introducing 3 products in 1 — youtube.com

When the three boxes coalesced into one, the crowd instantly realized the iPhone's revolutionary impact. Apple created the Swiss Army knife of personal digital devices. The introduction of the App Store and device connectivity evolved the iPhone into part of a robust ecosystem.

Apple's iPhone revolutionized expectations for phones, but it wasn't the first "smartphone". Most products or services replace or enhance already existing counterparts. Replacing established products or processes proves extremely challenging because many people resist change.

I imagined the potential for a graphical fixture offsets editor and later received patent 5985403 for the invention (Figure 64). These offsets simplified machining of multiple part blanks mounted to a rectangular fixture amusingly referred to as a "tombstone".

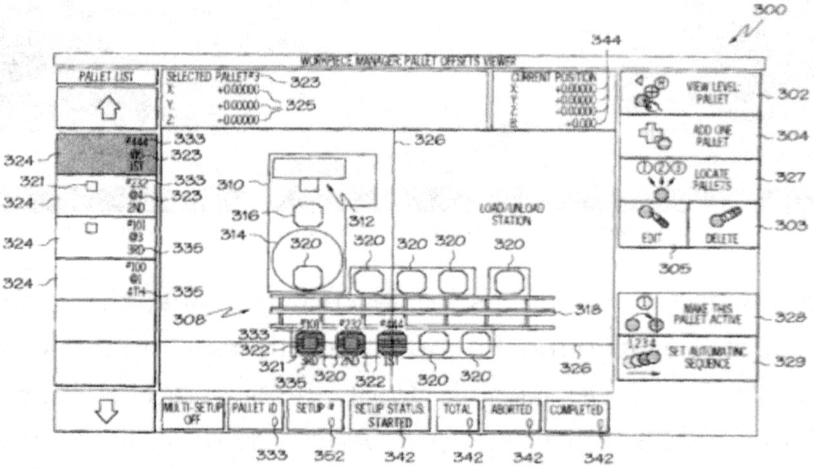

Figure 64:Workpiece Manager Patent — Justia.com USPTO

The previous process required tedious entry of data into various tables with no correlation to their real-world function. I will never forget proudly introducing my newly invented GUI for managing multiple fixtures to a crowd of expert machinists.

The demo went flawlessly. When I asked for questions, a grizzled veteran practiced with the existing painstaking process blurted out: "This is horrible". His reaction landed heavily on me. I would have liked to hear his feedback, but nobody else seemed to care. My colleague Dave found the incident amusing and repeated that line often.

I regret never getting to know my critic, but his reaction stuck with me for the rest of my career. Later reflection and empathy brought a realization: I took his special knowledge and replaced it with something that was easier, faster and did not require special skills. He realized I made his expertise available to less skilled machinists.

I hope he later recognized the true significance of his overall subject matter knowledge, but alas I will never know. My experience should prove cautionary to those under the spell of AI hype. Those wielding AI must remain in control and fully understand the strengths and weaknesses of non-human subject matter "experts". The path of blind AI dependence and obedience should alarm anyone considering it.

As another cautionary tale, I determined to buy Microsoft Flight Simulator 2024 (MSFS 2024) and even understood what happened as the developers rushed to ship a very ambitious product before the end of the year. It had 2024 in the name, right?. I gave up when realizing I was attempting to run a product that did not run well under the best of circumstances on my Mac using a Windows Virtual Machine (VM). My environment was neither one for which it was developed or tested. Attempting to wade through the overwhelming number of one star reviews squashed any remaining ambitions.

Figure 65:Microsoft Flight Simulator Crash — https://forums.flightsimulator.com/

Following introduction of Flight Simulator 2024, hordes of early adopters overloaded game servers for weeks. Intense hype surrounded the introduction and videos looked really amazing, but dashed hopes resulted in hundreds of dismal reviews. Experience teaches that the real world is not similarly kind to a software developer's carefully curated test or demo environment.

Prefer limited Alpha and Beta test programs to risky unlimited releases.

Another bleak example: Sonos replaced their well established and successful UI with something reviled for multiple reasons. This blunder resulted in firing CEO Patrick Spence even after his apology. Similar mistakes make recovery difficult to impossible because first impressions are everything.

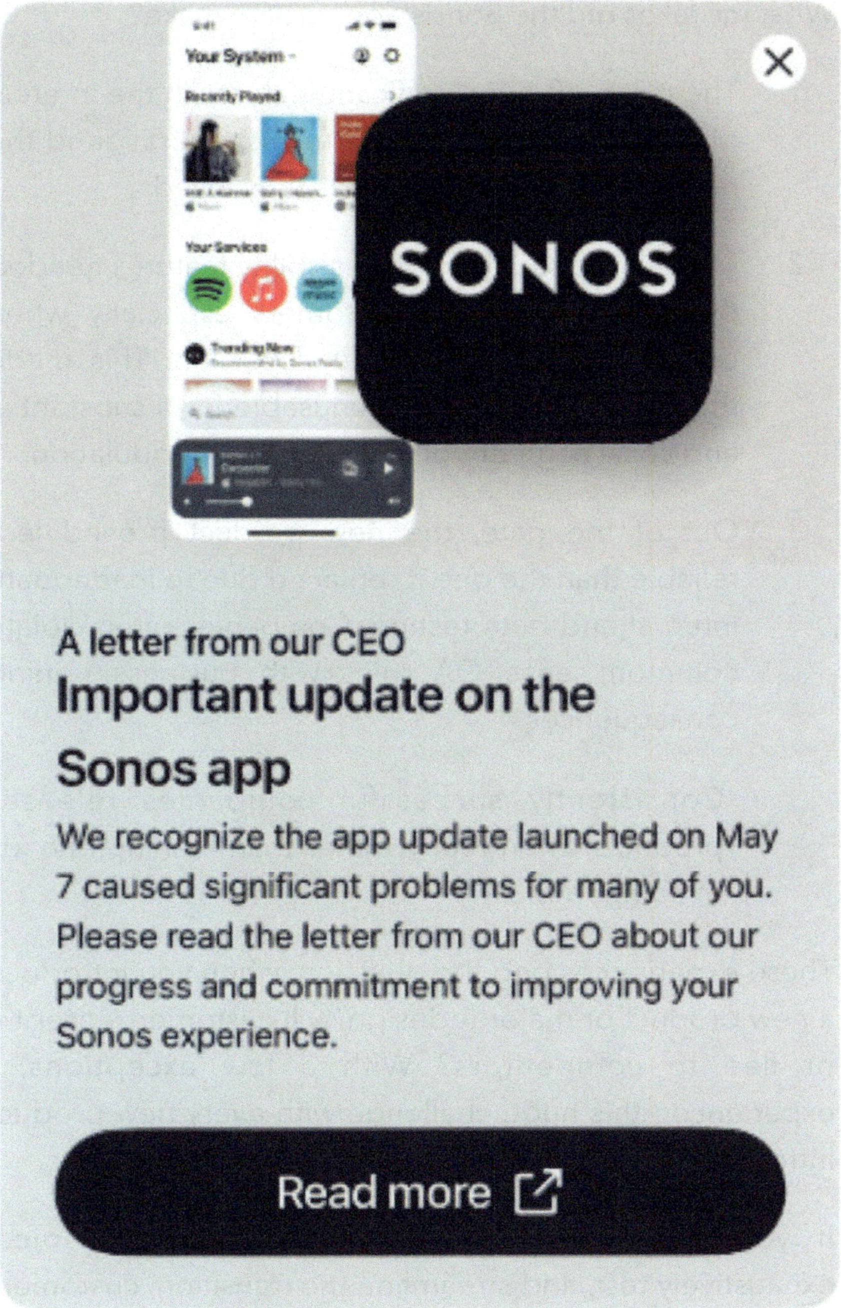

Figure 66: Sonos Apology - TheVerge.com

What mistakes did the Sonos app update make?

1. The team left out significant features in the interest of shipping earlier. Longterm customers found the new software lacked features they loved.

2. The release sacrificed accessibility features needed by visually impaired customers especially when compared with the replaced version. This made the new software simply unusable for a substantial and vocal segment of their customer population.

3. Out of the gate, the new product proved less reliable than the one it replaced due to inadequate internal and beta testing. Companies often violate common sense QA rules with business ruining consequences

Consistently successful companies release products when ready rather than rushing ahead of competitors.

These experiences beg the question, when you introduce a new product or major redesign, will customers accept it or flee to competitors? With a few exceptions, I experienced this huge challenge with every new product introduction.

If you don't give customers an upgrade choice, exhaustively test, and streamline the transition, customers may choose a competitor. After all, you've already forced customers to make a change. These examples demonstrate: the more drastic the change, the higher the risk.

There are other dangers forefront in our modern technology landscape. How do we cope with the frequent claims that AI technology will obsolete most of the workforce?

CHAPTER TWELVE

Living with the Promise and Problems of Artificial Intelligence

"Pretend Intelligence"
-- Richard Stallman, free software movement activist and programmer

Note that Richard Stallman's quote for this chapter applies to Large Language Models (LLMs), only one form of AI. However, there is an overall truth to this quote that we will examine. AI takes many forms, including expert systems, vision systems, symbolic AI etc. Many AI systems have subtle, but profound influence on our everyday lives including vehicle navigation, driver assistance systems, weather forecasting, pharmaceutical evaluation, etc. Please, do not interpret my statements that follow as generally "anti-AI".

Pundits tout LLMs and the possibility of general purpose AI as much more imminent and apocalyptic than indicated by reasonable analysis. However, we must now make efforts to rein in future applications that threaten our environment, security, and careers as explained here.

Let's consider the proper role of artificial intelligence in UX design going forward. In blunt terms, today's chatbots remind me of a throwback to 1983 (pre-Macintosh) command line interfaces.

People must frustratingly reverse-engineer the magic chatbot prompt to get a satisfactory response. Farming out computation to huge, energy hogging data centers only makes the situation worse. Many don't accept subsidizing all this infrastructure with privacy invasion for targeted marketing and also find yet another subscription model burdensome. Who wants to pay higher electric bills and increase fossil fuel burning to support routinely misused AI applications? In the past, forward thinkers foresaw better alternatives.

In 1987 Apple created the Knowledge Navigator video. I eagerly anticipated the introduction of intelligent assistants on call 24/7 to assist with daily activities. This experience required no arcane prompting and relied on natural back and forth discussion instead. Such interaction transcends today's chatbots. The device itself would locally handle the work. Almost 40 years later, release day for that product still has not arrived.

Figure 67: Apple's Knowledge Navigator concept decades before the iPad and Siri — reddit.com

The movie "Her" (2013), imagined an almost human-like AI companion. The movie eerily predicted intimate relationships with AI.

Figure 68: Samantha from the movie "Her". — BNONews.com

Some people see chatbots as more empathetic because they are programed to agree, and never to get bored or tired. Technology replacing human relationships rises above Don Norman's "Emotional Design" to be sure!

With the loneliness epidemic and aging population, AI may provide some measure of relief. However, real harm follows chatbots tendency toward approval. If a person considers self harm or harming others, the chatbot should disagree like a conscientious human would. Unfortunately there are documented cases of affirming such dangerous behavior (see Stanford Medicine/AI companions pose risk to kids). Companies like Anthropic employ philosophers to act as ethics guides to AI, but human intervention remains a better choice in risky situations for the foreseeable future (see article wsj.com Amanda Askell article).

Let's not ignore another vital consideration. Seamless integration requires AI to follow you all day, BUT should every event in a person's life be recorded and processed by an AI data center? For me and many others, privacy ranks high above the AI's fluency with my life especially if it increases privacy intrusion! If a company records increasingly personal details, potential buyers may sensibly avoid it and associated personal profiling that's already too common. Furthermore, others in your world may shun you when you wear a recording device as they did with Google Glass (even with indicator light).

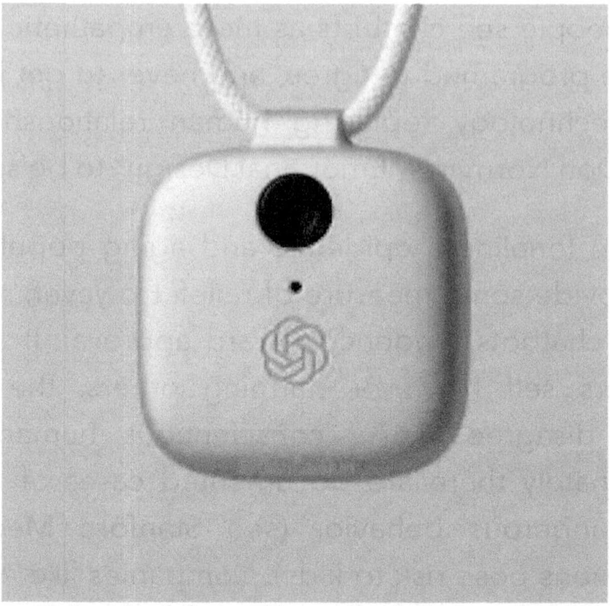

Figure 69: Imagined Open AI Jonny Ive device — cultofmac.com

With the rising importance of software and its corresponding complexity came the resolve of business leaders to find cheaper engineering alternatives. Countermeasures arrived in the form of offshoring development and its corresponding headaches including difficulties with communication and time zone incompatibilities. Another "solution" took the form of automation. Computer software writing software has taken many forms over the last 40 years without significantly impacting jobs. Companies only recently use AI as an excuse to significantly cut jobs and often later regret it.

An MIT report in 2025 found 95% of generative AI pilots failing at companies. Perhaps lessons learned will improve the efficacy, but this report counters much of the hype. Designers sensibly optimize AI for starting new products. Creating something new makes for great demos and easy test validation. Over time, and with more difficult problems, scaling issues rapidly manifest. Some LLMs seem adept at finding security exploits, but beware of a potential arms race between problems and solutions (nytimes.com).

Danger accompanies the misguided wish to have AI thinking for us. Research at the University of Pennsylvania determined that confident answers given by AI routinely lead people to accept results unquestioningly (Source: arstechnica.com/ai/2026/04 article: "Cognitive surrender" leads AI users to abandon logical thinking, research finds"). Time pressure worsens the effect. More wrong answers meant less trust (93% acceptance vs 80%), but still demonstrated alarmingly high acceptance without critical validation.

Rest assured, I wrote this book with very little AI assistance. I consulted Google with results now led with the Google Gemini AI. However, I validated such results with additional research, and critical thinking. I also tried some AI-assisted sentence rewrites, but almost always thought: "That's not what I meant!"

The more critical your plans for AI results, the more critical validation you need.

Dependency on AI may result from low self esteem or laziness, but acting on bad advice often ends in tragedy. Early AI gave a confidence factor, but instead, current AI routinely gives wrong answers in confident terms. Unlike people, AI doesn't indulge in second thoughts. As of early 2026, multiple instances of inadequately vetted AI generated code passed into system use, resulting in system outages. Having AI agents act on our behalf must provide stringent guardrails, especially for destructive actions.

Companies like OpenAI and Anthropic create most of the Large Language Models using a process akin to trial and error. Human programmers build algorithms that build neural networks (like connections between brain cells). Huge server networks run the algorithms on huge datasets to "train" the LLM. Finally, humans use feedback to "fine-tune" the model. (Paraphrased and summarized from Google Gemini, of course).

Unlike a conventional computer program, there is no code to read and validate. How do you determine the quality and completeness of the training source material? When the source material conflicts, how are the conflicts resolved? Nobody fully "understands" exactly how LLMs arrive at their responses though chatbots provide some chain of reasoning when asked. Even worse, the results are unpredictable.

Trusting AI requires understanding its decision making process, especially for agents.

We already infuse weapons with AI agency at our peril. Ukraine used an AI vision system to target Russian bombers in 2023 (in operation Spiderweb). Therefore, we increasingly find ourselves on the slippery slope from hunting planes with AI weaponry to autonomous people hunting robots. In March 2026, the Pentagon's war department made attempts to coerce Anthropic to comply with inadequate guardrails that were thankfully rebuffed, so far (source: bbc.com March 26, 2026: "Judge rejects Pentagon's attempt to 'cripple' Anthropic").

For almost 60 years, many movies like "2001 A Space Odyssey", "The Terminator", "Ex Machina", and more, educated us about potential dangers of autonomous robots. Aside movies, numerous scholarly papers, talks and books amply forewarn of consequences stemming not from AI's malice, but indifference. Anyone advocating autonomous AI weaponry must consider such warnings thoroughly and be held publicly accountable for their actions.

Why should we trust technology that has no real stake in the outcome? Does Alan Turing's test requirement of fooling humans into thinking AI is a human matter anymore? In business, humans still excel at valued skills of face-to-face interaction.

People need to forge trust relationships based on authenticity and with a person holding accountability.

Our success as a society comes from effective collaboration between people first and tools later. For the foreseeable future, knowing how to use the right tool for the right job in the right way remains uniquely human.

As of this writing, AI may play a role in your user experience design, but only humans understand multifaceted decision making and nuances of AI model applicability to a given situation. Lacking sentience, artificial intelligence imagines nothing. It has no true appreciation of people as you or your team do. At best, AI reformulates existing human wisdom, acts as a tireless sentinel, absolves us of tedium, or repeatedly tries multitude alternatives until it arrives at a solution.

Most public attention focuses on AI involving chatbots. What new vistas do AI vision systems and potent microprocessors allow us to envision?

CHAPTER THIRTEEN

Envisioning Vital Details with Augmented Reality

"AR allows individuals to be present in the world but hopefully allows an improvement on what's happening presently. Most people don't want to lock themselves out from the world..."
– Tim Cook, CEO of Apple

Artificial Intelligence (AI) and Augmented Reality (AR) offer modern inflection points like the GUI inflection point in my early career. Those who look at technology systems and envision futures building on new and evolving technologies will enjoy exciting new opportunities. After numerous false starts, better assistants appear on the horizon. Products like glasses and cars already feature AR, but such products continue to evolve, similar to early GUIs. Some of these technologies may trigger a vision of the future for you like the Mac did for me.

Xerox Palo Alto Research Center (PARC) invented the technology combining a mouse driven interface and over 300,000 pixels on display. On January 19, 1983 Apple introduced the Lisa computer based on seeing a demo of PARC's technology. Lisa pioneered Apple's GUI technology ahead of the Macintosh. While still in college, I acted as computer consultant to my father's small real estate business. In spite of my fascination with Lisa's breakthrough technology, dad wisely passed on buying one. However, only a few years later, he and his staff used a Macintosh and LaserWriter printer to generate top quality, flawless appraisal reports for his clients. The professionalism of no "white out" mirrored his well known attention to detail, earning him a solid competitive advantage.

Figure 70: Apple's Lisa: pioneering GUI computer — macstories.com

Apple's Vision Pro has much in common with Apple's Lisa. Vision Pro transcends conventional VR with "spatial computing" and pioneers augmented reality for anticipated Apple AR glasses. Both Lisa and Vision Pro were ahead of their time, but held back by being expensive and difficult to justify for daily use.

Figure 71: Vision Pro spatial computer: Apple's pioneering VR + AR device — apple.com

Vision Pro spatial computing allows people to place objects persistently in their environment and run multiple virtual applications. Apple also developed a powerful vocabulary of simple hand gestures for manipulating virtual objects.

The heavy Vision Pro headset masquerades as goggles, but observers only see imitations of the wearers "eyes".

Depending on immersion setting, people may experience Vision Pro as AR, but otherwise the product has more in common with VR. (Figure 72)

Figure 72: Vision Pro partial immersion — apple.com

Below, Vision Pro adds widgets, more like AR glasses, but places widgets spatially somewhere in the room rather than statically positioned in the lenses of the glasses.

Figure 73: Vision Pro showing AR widgets — roadtovr.com

People experience Virtual Reality (VR) much differently than AR because VR goggles still enclose and isolate people inside a separate environment. Isolation works well for immersive games, simulation and movies, but so far such enclosure discourages wide, lifestyle-altering, adoption. Everyday activities require face to face interaction and long term, unencumbered wearing.

In contrast, lightweight AR glasses and AR Head Up Displays overlay only what people normally see. Observers may be unaware of augmented experiences presented by AR glasses or a car's HUD. See Figures below:

Figure 74: AR Glasses with Overlaid Info — allaboutvision.com

Figure 75: Panasonic AR HUD with Overlaid Info — autovista24.autovistagroup.com

Early research into adopting AR already demonstrates its benefits. For example, the National Institutes of Health

(NIH) focused on treating phobias and PTSD using AR-enabled safe and gradual exposure therapy (pmc.ncbi.nlm.nih.gov PMID: 39282478). From a different perspective, Florida Atlantic University improved training outcomes when using AR to guide people with developmental and intellectual disabilities (inc.com – Kit Eaton).

With augmented reality, people focus outward rather than downward at fixed screens.

AI helps interpret and annotate what we see and hear while AR keeps us immersed in the present moment. In my experience, AR reintegrates us with our environment and allows us to focus on what's ahead.

Drivers may choose from voice prompting and/or Head Up Displays rather than distracting dashboard screens. Obviously, AR designers should avoid overlaying critical areas of the visual field, and also avoid distracting elements that continually update or flash.

AR technology for cars already demonstrates great promise, such as the following example. The Porsche HUD features always available speeds and speed limits. An adjustment keeps them just out of the direct line of sight but without requiring refocusing. Other information appears as necessary. For example, upcoming turns and optimal exit lanes appear as needed. Settings such as parallel lines for cruise control's "following distance" appear when the driver adjusts a switch. The white and green distance lines disappear shortly after (see below). Technologies such as AR improve driver confidence and invigorate road trips.

Going forward, including such augmented reality features promote better driver focus and awareness. (See Figure 76)

Figure 76: Augmented Reality Head Up Display - porsche.com

The best AR fuses decades of AI vision system development with aspirin-sized chips teeming with neural engines, graphics cores and RISC supercomputers to conjure up an enhanced view of the world (Figure 77).

Figure 77: Apple M5 Chip with 10 Core CPU & GPU, 16 Core Neural Engine - apple.com

I see vast potential here for AR improved situational awareness that you and many following generations may realize. Just a few examples for AR glasses:

- Staying in touch: AR displays your text messages, email headers and vital news. Does it supplement or replace smart watches?

- Music appreciation: AI analyzes ambient sounds and AR displays the song title it hears in the background along with lyrics if preferred.

- Hiking assistance: AR warns of hazards ahead such as steep drop-offs and swamps. AR identifies plants and animals and highlights poisonous varieties along with other risks. AR displays directions to get you back to your car or home.

- In the supermarket: AR navigates you through your grocery list for rapid pickup and points out specials of interest to you.

- Enhanced telemedicine: Doctors annotate concerning conditions including rashes, skin growths and wounds. Doctors may point out potential risks and disease indicators.

- Astronomical observation: AR identifies constellations, planets and other celestial bodies.

Surveillance concerns doomed previous glasses. Now, vendors must establish trust in both wearer and onlooker. To help accomplish this, newer AR glasses should look more like typical fashionable glasses. On introduction, product education and news coverage make or break public acceptance. Allowing any recording may require gesture cues to onlookers or some other trustworthy action. Like many, I don't trust always-on recording devices.

After considering the current and future promise of Augmented Reality, we consider an imaginary future product. Assembling all the puzzle pieces from previous chapters reveals guidelines for evaluating and building future products.

CHAPTER FOURTEEN

Riding Current and Future Technology Waves

"The best way to predict the future is to invent it"
– Alan Kay, pioneered object oriented programming and GUI at Xerox PARC. Proposed the Dynabook tablet 50 years before iPad.

When your user interface elevates daily life for its audience, it redefines the state of the art. Let's summarize central ideas in this book into 7 guidelines that make or break your UI. We then examine a future project relative to our guidelines.

I've had experience with numerous challenging goals that required a longterm vision.

When I was still in college, serving as a cooperative student (now called "intern"), I imagined and printed a crude dot-matrix screen layout of a graphical computer machine control (CNC). I then delivered the image to the research & development (R&D) department. I'm sure they laughed at the time, but I ended up on a team 10 years later that built a touch-screen-driven GUI concept from the ground up. (See Figure 4) The trip to the machine tool show to demo our state-of-the-art creation rewarded many hours of pre-show overtime.

Many years later and spanning almost another decade, I dreamed of dashboards for material handling systems. Our team, later designed and implemented dashboards having similar features. Those streamlined, powerful and user-configurable dashboards demoed well to customers. Better yet, they exceeded anything I had imagined.

Before we explore a future idea, here are 7 guidelines summarizing the previous chapters we will use to conceptualize and evaluate a hypothetical example below. I base the guidelines on hard lessons of my own. These guidelines differ significantly from but include Don Norman's (Chapter 2) such as being:

1. **Appropriately organized,** according to your audience. You should categorize UI navigation according to a typical operator's understanding of real-world objects (mental model). Software catered to a variety of workplaces may lack a "typical" customer. For more general audiences, create a configurable "Home" screen with sensible defaults. People optionally choose to tweak your standard to better suit their needs.

2. **Familiar,** by applying standard patterns and not reinventing without major benefit. Layout widgets either proximate to devices they control or in start up sequence order. When in doubt, let people customize. Interaction should follow typical gestures or patterns of use.

3. **Helpful,** by anticipating each step to enhance a person''s workflow and providing great defaults (maybe even provided by AI). This means avoiding anti patterns and dark patterns, too!

4. **Beautifully elegant** by combining aesthetics and simplicity. Design teams provide a potential head start or at least challenge your perspective. However, capture proper layout rules in the framework such that the implementation team can't help but get it right. You succeed by transforming complexity into simplicity. (As an unfortunate side-effect, colleagues think your job is easy, but people underestimate masterpieces of intuition because they *seem obvious* in hindsight).

5. **Concise**, rather than an obstacle to productivity, on screen text and dialogs benefit from bite-sized wording. Collaborating with the documentation team improves not only on screen text but also the workflow of the UI itself.

6. **Forgiving**, by providing undo or second thought verification of risky actions and explaining consequences. The tricky balance between security and ease of access also applies. Cutting corners, such as cars making grating noises when approaching obstacles (especially when they can safely move around them), sabotages otherwise great products and encourages people to rid

themselves of their annoyances. Keep humane treatment of error conditions carefully in mind.

7. **Responsive**, by exceeding performance requirements. Avoid customer and support hardships by providing adequate CPU bandwidth, memory capacity, storage space and stress testing to anticipate adverse conditions. Demand attention when resources run low. Include appropriate contact information whether it be customer local or to your company's tech support.

As I wrapped up this book, I composed the cover art including an imagined augmented reality picture window to provide a fun and forward looking example. The visible elements represent effective UI at its best. AI detects and identifies the white eagle and AR visually annotates it. AR annotations come and go as befitting the moment.

AI processes the scenery and determines what's changing and potentially interesting while filtering out noisy elements to keep augments useful and minimal. The window's owner might customize the window to include weather with air and water temperatures, wind speeds and other interesting details.

Figure 78: Augmented Reality Picture Window with AI - Author & John Ruthven

These systems enhance our experience by keeping us alert and aware. They may detect people, animals, weather, astronomical phenomena and perhaps something unexpected. The system processes images locally on the device with maybe occasional exceptions for visuals not immediately identifiable. The system may store some of these images locally allowing people to decide whether they warrant further investigation. Importantly, these devices should not contribute to a surveillance network, such as the one proposed by an infamous 2026 Super Bowl ad.

Additionally, some propose scanning the skies hoping to capture and identify so called "Unidentified Aerial Phenomena". Should such an occurrence happen near my home, I would be elated to have AI capture and record what I might otherwise miss. You can find research about scanning for UAP in the sciencedirect.com article by physicist Matthew Szydagis.

Let's take the previously mentioned guidelines and relate them to our AR + AI picture window example. We then observe it's alignment with the 7 guidelines:

1. **Appropriately organized:** AR information overlays pictured scenery proximate to where it appears. Water and air temperatures overlay their respective environmental imagery. Annotations appear conveniently juxtaposed only to changing scenery unless otherwise specified.

2. **Familiar:** AR may reuse familiar icons and idioms from GUIs like the thermometer and warning icons while inventing new presentations that better overlay their surroundings. Air temperature might annotate the sun's location if shown in frame rather than the sun icon added in the example .Figure 76

includes both familiar icons and image overlays similar to recent weather apps.

3. **Helpful:** AI learns its human counterpart and provides personally tailored information. AI also recognizes and adapts to visual changes in its environment. The device supports scanning for otherwise missed interesting phenomena while avoiding undesirable "big brother" type networked surveillance. Privacy and integrity play a huge role in product acceptance.

4. **Beautifully elegant:** AR integrates with the scene's surroundings such as adjusting to seasonal views and to weather and light levels. Weather icons and imagery complement the view rather than conflict with it.

5. **Concise**: Since AR overlays environmental images, people infer meaning by visual context. Minimal wording keeps annotations concise avoiding encyclopedic information (unless specifically requested). Minimal annotations avoid visual clutter.

6. **Forgiving:** settings for the AR window include online backup. Restore to Factory Defaults has an Undo feature or requires confirmation. The device keeps locally stored captured images even if deleted unless storage runs low and the device

deletes oldest first.

7. **Responsive:** Information updates either in realtime (for fleeting situations) and/or includes some sort of replay or persistence feature. Your company supplies hardware that presumably guarantees adequate resource headroom. Alternatively, an installer validates hardware requirements.

With summary guidelines in place let's reveal the purpose and destination of our journey.

CHAPTER FIFTEEN

Making Your Unique Contribution

"What you do makes a difference, and you have to decide what kind of difference you want to make"
– Jane Goodall, Primatologist, Conservationist and Anthropologist

Great ideas turn into great products only with Herculean effort. This means concentrating on all details from crisp graphics, concise meaningful text, pertinent data, and solid supporting infrastructure. No UI stands alone. It must be part of an overall well engineered system. The UI must also access high quality data and metadata.

On top of all the technology details, organizational bureaucracy complicates UI decisions. I have worked with colleagues who expertly and patiently handle such issues, including my most recent manager. However, leadership needs support of the entire team to fairly represent the UI. My early experience prepared me to address and support impactful UI discussions.

I was fortunate to join a cross-disciplinary lunch group in my early career. It also featured a set of widely different points of view across the political spectrum from Libertarian, Democratic and Republican registered voters. We worked for different managers and held different specialties. The conversation helped us stay knowledgeable about how projects were progressing across departments and allowed us to answer questions that made each of us more effective in our jobs. More importantly, it taught me what it means to be a "voice of reason". We were each able to listen earnestly and speak respectfully. We solved engineering challenges, pointed good natured fun at our own politics and above all, exercised humility.

Such conversations have been largely replaced by social media, but the format and algorithms isolate people with haphazard and disjointed messages. Social media cannot match face-to-face meetings.

Try forming a cross-discipline colleague lunch group, even if only on Zoom or FaceTime. Practice being a voice of reason at work or otherwise.

Because the UI represents the system to its audience, people sometimes see it as responsible for software and hardware shortcomings found elsewhere in the system, as discussed in Chapter 8: "Avoiding Disasters". Customers also specify alternative UIs. For these reasons and others, colleagues enter into highly charged discussions about replacing the UI that trivialize the difficulty and consequences. Those advocating replacement tend to vastly underestimate associated time and effort.

Discussions about UI challenge our ability to listen to and consider all viewpoints compassionately. Unlike some current social norms, pursue doing the right thing as the goal rather than "winning." If, after careful consideration you disagree, do so patiently and from the point of view of your customer, your business and with solid reasoning. At best, a UI provides a cohesive view and expertly unifies all parts of a system in the service of everyday people.

If you favor a voice of reason in discussions, and your leaders recognize this, you may be recognized such as in the insight below:

> "You have the temperament for the UI" means you have a customer perspective and consider disagreements with wisdom and grace.

Walter Isaacson's *The Innovators* cites extraordinary advances of teams at Apple, Microsoft, AOL and many others. Isaacson realizes the common thread that technology succeeds by enhancing human potential. Sharing Isaacson's insight rewards a career building compelling user interfaces as it's vital heart. With all of the technology trends and upheaval, your own meaningful contribution to enhancing human potential still calls out to you.

Figure 79: Walter Isaacson's book *The Innovators* — Amazon.com

Opportunities appear unexpectedly from novel combinations of your expertise and that of your colleagues, because as Leonardo da Vinci said: "Realize that everything connects to everything else". Da Vinci synthesized his vast contributions from a combination of art, anatomy, and engineering.

As another example, Steve Jobs studied calligraphy in college and later insisted on a variety of fonts for the Macintosh. Mac typography might otherwise be bland were it not for his experience. Jobs also mastered envisioning the future. His visions lacked detail, but he surrounded himself with people that brought those ideas into focus and full-color clarity. Jobs also smartly tackled projects that were merely improbable. Some of Jobs' modern contemporaries instead promise the impossible.

Your background, aptitudes and talents uniquely qualify you to achieve greatness that's unique to you. Just as I learned from insightful authors, colleague subject matter experts, and attending conferences, continual enrichment also beckons you. Ultimately, you and your team's mastery of technology, imagination and relentless drive creates mutually beneficial change that improves our lives.

Even decades later, you may discover and enjoy the continued relevance of your work. Recently, I found and watched a 2023 YouTube video demonstrating how to use the multi-setup feature of the GUI machine control, from early in my career. The skilled machinist deftly works around a few user experience obstacles that I would ideally address. However, our team's machine control still works remarkably well compared to many 30 year old technologies. Here is the expanded video link from the preface: www.youtube.com/watch?v=zoyfb12LQG0.

Personal mastery (and fulfillment) also comes from passing along guidelines and stories. As my career wrapped up, I focused on mentorship. Highlights of my experience and career journey appears in this book. As usual, our team worked on both a new UI and maintained a legacy UI. Many team members preferred to concentrate on the new while ignoring the old, but both hold equal importance to me. My advice: When you are working on the new UI, love the new UI. When you are working on the legacy UI, love the legacy UI. (Clearly that applies not only to UI, but many other worthwhile endeavors...) Even seemingly mundane aspects of the UI benefit from care and concern!

Good UIs create a feeling of home that arises from the love that designers and developers gave them.

A few final thoughts... Approach each project as an opportunity to make life better and communicate with a voice of reason, understanding and awareness. Such genuinely human traits elevate you above competition (including AI) and maximize your value and personal satisfaction. Keep a wistful minds eye on visions of a better future, especially when it seems most unlikely. Information in this book provides a recipe, but your vision and skills supply the essential ingredient going forward. Whether figurative or literal, your moonshot awaits.

I wish you greatest success improving our world and humbly realize that your caring actions make all the difference for our future. Create compassionately!

About the cover

The cover composites the past, present and (potential) future of UI design.

A picture found on Wikipedia of the original Macintosh represents the past. It was edited to remove the background and floppy drive for aesthetic purposes.

A photo of my iPad displaying the Apple Home app represents the present, along with a monitor displaying KPIs from Kyubit: https://www.kyubit.com/kpi-dashboard

A conceptual design of my own for an augmented reality smart window imagines the future. To my knowledge, there is no product like this, but most of the technology exists.

The scene includes an image from StockCake.com https://stockcake.com/i/snowy-lakeside-morning_1318980_875533. The sun icon is from a weather app image here: https://stock.adobe.com/images/weather-forecast-widget-at-smartphone-screen-vector-illustration-mobile-phone-with-daily-weather-forecast-application-template-sun-clouds-thunderstorm-and-rain/385336798 and the water temp thermometer is from https://www.istockphoto.com/illustrations/thermometer-icon. The caution icon is from https://emojipedia.org/warning. The White Eagle is from a John Ruthven print entitled "White Eagle (Aguila Alba).

The eagle Ruthven painted is a rare partially albino golden eagle. You can read more about it here https://www.audubon.org/news/rare-white-eagle-stirs-souls and you'll find more in Google with this search string "white eagle spiritual meaning" lest I spoil it here. Some find the concept of a totem animal to be a bit quirky, but if there ever was a totem representing my aspirations, this is the one.

Made in the USA
Monee, IL
03 November 2026

67384329R00095